Arming Asia

Bitzinger examines the phenomenon of attempted self-reliance in arms production within Asia and assesses the extent of success in balancing this independence with the growing requirements of next-generation weapons systems. He analyzes China, India, Japan, South Korea, Southeast Asia, and Taiwan.

The overarching question in the book is whether self-reliance is a strategically viable solution for the development and manufacturing of arms. Given the ever-changing dynamics of armaments production and the increasing demand for sophisticated next-generation weaponry, will these countries be able to individually sustain their domestic defense industries and constantly update their technologies?

This is one of the few books to analyze Asian arms production from a regional perspective.

Richard A. Bitzinger is a Senior Fellow at the S. Rajaratnam School of International Studies, Nanyang Technological University in Singapore. He is the author of *Towards a Brave New Arms Industry?* (Oxford University Press, 2003), "Come the Revolution: Transforming the Asia-Pacific's Militaries," *Naval War College Review* (Fall, 2005), *Transforming the U.S. Military: Implications for the Asia-Pacific* (ASPI, December 2006), and "Military Modernization in the Asia-Pacific: Assessing New Capabilities," *Asia's Rising Power* (NBR, 2010). He is also the editor of *The Modern Defense Industry: Political, Economic and Technological Issues* (Praeger, 2009) and *Emerging Critical Technologies and Their Impact on Asian-Pacific Security* (Palgrave, 2016).

Routledge Security in Asia

Arming Asia

Technonationalism and its impact on local defense industries

Richard A. Bitzinger

LONDON AND NEW YORK

First published 2017 by Routledge

2 Park Square, Milton Park, Abingdon, Oxfordshire OX14 4RN
711 Third Avenue, New York, NY 10017

Routledge is an imprint of the Taylor & Francis Group, an informa business

First issued in paperback 2018

British Library Cataloguing in Publication Data
A catalogue record for this book is available from the British Library

Library of Congress Cataloging in Publication Data
A catalog record for this book has been requested

ISBN: 978-1-138-89255-2 (hbk)
ISBN: 978-1-138-36602-2 (pbk)

Typeset in Galliard
by Apex CoVantage, LLC

For my daughters, Amy and Jennifer

Contents

Acknowledgements

This book could not have been possible without the considerable support and inputs of many people. In the first place, I am grateful to the management at the S.Rajaratnam School of International Studies (RSIS) – particularly Executive Deputy Chairman Ambassador Ong Keng Yong, Dean Joseph Liow, and Dr. Tan See Seng, Director and Head of Research at RSIS's Institute of Defense and Strategic Studies – who gave this book project their full and enthusiastic backing. I am also grateful to my many friends and family who contributed advice, succor, and support during the long and often arduous process of writing this book.

R.A.B., Singapore, November 2016

Introduction

Asia is often overlooked as a center of armaments production. Most of the largest arms-producing states are still located in North America and Europe, and as a whole the Asia-Pacific region probably accounts for less than ten percent of all global weapons manufacturing. Nevertheless, that still makes it the world's third largest defense-industrial hub – and it is growing, in terms of the numbers, variety, and the sophistication of its weapons systems.[1] Moreover, as defense spending continues to languish in Europe, the focal point of global armaments production is slowly shifting from the North Atlantic to the Asia-Pacific. Some of the world's biggest military spenders are located in Asia, including China, India, Japan, Pakistan, and the Republic of Korea (South Korea). China, in fact, already has the world's second highest defense budget; its estimated military research and development (R&D) spending (perhaps US$10 billion) is twice as great as all of Europe combined.[2] And while it is true that Asian-Pacific militaries still import large chunks of their arms from the West, this trend will not last. Most of the biggest military spenders in the Asia-Pacific also possess sizable defense industries, and their governments are committed to increasing their purchases from local arms suppliers.

Nation-states – not just in the Asia-Pacific but around the world – have many reasons to produce armaments, but traditionally the strongest motivation has been classically realist and security oriented: the need to provide for a secure source of military materiel necessary to deter threats and to defend one's national territory. Possessing or attempting to possess strong domestic arms industries, capable of designing, developing, and manufacturing advanced weapons systems, is viewed by many countries as an essential element of this strategy. Consequently, *autarky*, or self-sufficiency in arms acquisition, can be a critical national security objective. At the same time, however, such autarky traditionally had quite limited military motivations, i.e., national defense. Increasingly, however, many nations – and particularly those in the Asia-Pacific – have come to view indigenous arms production from a much broader perspective: the idea that autarky in armaments serves larger, more ambitious national interests, i.e., it is about securing and advancing a nation's geopolitical status in a regional or global system. This particularly *technonationalist* approach to armaments production has become endemic to the Asia-Pacific region, and it is critical to understand why and how this trait has so strongly influenced regional defense industrialization and arms manufacturing. It is also important to always keep the "technonationalist impulse" in mind when addressing how Asian-Pacific nations deal with problems and failures when it comes to indigenous armaments production, and why, despite whatever setbacks they may encounter, maintaining and expanding their national defense industrial bases remains a high priority.

The Asian-Pacific defense industry is unique in its persistence in following a decidedly technonationalist approach demanding self-sufficiency in armaments production. Among the nations in the region who do produce arms, there is an almost obsessive predilection for self-reliance among the states in this region when it comes to developing and manufacturing arms, and consequently these countries have invested considerable resources into their defense technological and industrial bases. At the same time, technonationalism in armaments production is not an easy course; in fact, it has often been deceptively difficult. Nevertheless, in spite of these challenges the siren song of military technonationalism remains a powerful call. What is the most interesting about Asian-Pacific arms industries is how enduring they have been *despite* the fact that they seem to produce few economic benefits or, more critically, to contribute so little (relative to their costs) to expanding a country's military-technological capacities for national defense. Asian armaments production has rarely been cost effective or militarily significant in terms of turning out state-of-the-art military equipment. And yet, most large powers in Asia have not abandoned the idea of autarky in armaments production. On the contrary, China and India still have *explicit* technonationalist industrial strategies when it comes to arms acquisitions, and the pattern of defense industrialization in South Korea over the past 25 years shows that autarky still wields strong influence there as well. Even Japan is not giving up on certain dreams of self-sufficiency, as evidenced by its recent decision to greenlight the ATD-X stealth fighter program (and which had it first flight in early 2016). If Asian-Pacific arms producers are persisting in their efforts despite the costs or dubious military benefits, then it is probably more due to the technonationalistic appeal of autarky than any to economic or national-security gains.

The challenge to Asian-Pacific arms industries is meeting the growing demand for self-sufficiency in arms acquisition, i.e., autarky in production, as well as the rapidly increasing technological requirements of next-generation weapons systems. In other words, can Asian defense factories develop and produce the types of advanced weaponry that their militaries increasingly clamor for, and do so under domestic political and economic conditions that demand increasing self-reliance in production, from initial design all the way to final manufacturing? The dilemma facing these countries is whether such a go-it-alone strategy is still feasible – that is, can it build and sustain technologically advanced domestic defense industries? In other words, even if these countries are willing to pay the "technonationalist" premium for continued autarky, will it be sufficient for the task of developing and manufacturing next-generation weapons systems?

This book addresses the rise of defense industries and the process of indigenizing armaments production within several newly industrialized states in the Asia-Pacific region. In particular, it examines the process of defense industrialization among the leading arms-producing states in the Asia-Pacific and how "military technonationalism" has not only driven defense industrialization in the region, but also how such technonationalism has also provided a model for development (i.e., with the ultimate objective of autarky). It explores the paradoxically symbiotic relationship between technonationalism and "technoglobalism," and the critical role that foreign technologies have played in process of defense-industrial indigenization. Finally, it discusses whether the technonationalist model is a viable or sustainable approach in terms of economics and (especially) military innovation. When we address the experiences of individual Asian nations as they have tried to apply the technonationalist model to their respective defense industries, we will constantly return to many key questions, including: has technonationalism been a successful strategy for Asian-Pacific efforts at defense industrialization? Where has it succeeded – or failed – and why? Does indigenous armaments production – as expensive and technologically demanding as it often is – make sense for some countries, especially smaller states? If not, are there alternative paths, and what are their prospects? In conclusion, this book argues that despite the disincentives surrounding indigenous armaments production – in terms of the high cost of autarky and the dubious military gains that tend to accrue – most Asian states will not abandon their defense industries or the goal of achieving autarky, and that is due mainly (and increasingly) to the driving force of military technonationalism. Despite everything, it is still difficult to separate autarky from technonationalism, and technonationalism from autarky.

Notes

1 Extrapolated from Elisabeth Sköns and Eamon Surry, "Arms Production," in *SIPRI Yearbook 2007* (Oxford: Oxford University Press, 2007), pp. 345, 347–48.
2 According to the European Defense Agency, total EU military R&D spending was around €5 billion (US$6.3 billion) in 2012, down by nearly half from 2006 (€9.8 billion, or US$12.4 billion) (http://www.eda.europa.eu/docs/default-source/eda-publications/defence-data-booklet-2012-web).

1 Military technonationalism and the drive for indigenous arms manufacturing

Nearly every country in the Asia-Pacific region manufactures some kind of arms, but the type, quantity, and quality of weaponry produced vary widely. In some cases, armaments production is relatively inconsequential, consigned to relatively "low-tech" types of weapons systems, i.e., small arms (rifles and pistols), ammunition, armored cars, aluminum-hulled patrol boats, and the like. Moreover, much of this manufacturing is based on the assembly of foreign weapons systems, such as the licensed-production of US M16 assault rifles or the putting together of imported kits for helicopters or armored vehicles. Consequently, most Asian-Pacific arms manufacturers could at best be described as "third-tier" producer-states, occupying the lowest strata in the global hierarchy of arms-producing states. That said, there exist a few countries in the region – so-called "second-tier" producer-states – that are heavily (or increasingly) engaged in arms manufacturing and which have built up quite extensive, and in some areas quite sophisticated, indigenous arms industries, capable of designing, developing, and manufacturing their own weapons systems.[1] These are generally of three types: (1) technologically advanced industrialized countries that produce a limited array of state-of-the-art weapons systems (e.g., Japan); (2) newly industrialized economies containing modest (but growing, both in terms of range of production and military capabilities) military-industrial complexes (e.g., Indonesia, Singapore, South Korea, and Taiwan); and (3) rising great powers with large, broad-based defense industries, but which are still lacking in certain areas of indigenous research and development (R&D) and industrial capacities when it comes to developing and producing highly sophisticated conventional arms (e.g., China and India).[2] It is this rather catholic grouping of second-tier Asian-Pacific producer-states that we are most concerned with in this book.

Despite their varied backgrounds and varying defense-industrial capacities, these countries share many motives for developing and producing their own arms. One of the strongest of these is the security-driven imperative for self-reliance in defense.[3] In a basically anarchic international security system, nation-states are naturally impelled to seek an independent defense capability. In order to defend its territory satisfactorily, therefore, a nation-state requires a reliable source of armaments, and the most dependable source is generally a domestic one. Additionally, relying too heavily on arms imports means exposing the nation-state to embargoes or to technology holdbacks, thus risking its ability to acquire the weapons it deems essential to its national defense. Such foreign dependencies can also leave a country vulnerable to coercive efforts by a supplier-state who might try to use the threat of withholding arms deliveries in order to extract certain kinds of behavior on the part of the recipient, such as respect for human rights or stopping international aggression by the buyer-state. In a global environment where supplier restraint – whether real or potential – is a critical national-security concern, security of supply can be a key factor driving local defense industrialization.

In addition to fulfilling perceived requirements for self-sufficiency, arms production has often been seen as an important mechanism for driving a country's overall economic development and industrialization. Defense industrialization had potential backward linkages spurring the expansion and modernization of other sectors of the national economy, such as steel, machine tools, and shipbuilding.[4] Industrialization and technological advancement was seen as feeding into the development of domestic arms-manufacturing capabilities, such as building up general skills and know-how and providing lead-in support or equipment for arms production. The construction of warships, for example, stimulated the establishment of indigenous shipbuilding industries, for example, while production of military vehicles required steel mills and automotive factories to provide critical parts and components, such as armor plating, chassis, and engines, and skilled labor to assemble these vehicles. The Republic of Korea (ROK), for example, consciously pursued parallel strategies of "security and development," that is, building up its heavy industry and high-technology sectors at the same time as it strove for self-sufficiency in arms production.[5] Particularly during the Park Chung-hee regime (1961–1979), South Korea's leadership viewed economic development to be an essential element of national security, and as such, state investments in heavy industry sectors like steel, machinery, transportation, and chemicals were regarded as directly contributing to defense preparedness.[6] Industrialization and technological advancement were seen as feeding into the development of domestic arms-manufacturing capabilities, such as building up general skills and know-how, and providing lead-in support or equipment for arms production. The establishment of the South Korean commercial shipbuilding industry, for example, facilitated the construction of warships, while the creation of a domestic steel industry and, later, a domestic automobile industry, provided the parts and materiel (such as armor plating, chassis, and engines), as well as the skilled labor necessary for the production of military vehicles.

At the same time, armaments production was viewed as a "technology locomotive" spurring the growth of new industries and new technologies, particularly in the area of aerospace, electronics, and information technologies sectors.[7] Military aerospace programs, for example, often constituted the basis for civil aircraft and aviation production in nearly all of the second-tier arms-producing states. Initially based on military-led industrialization, for example, Brazil's Embraer subsequently expanded into the regional jet business; Israel's various high-technology sectors have also benefited greatly from cross-fertilization with military industries.[8] For its part, South Korea has attempted to exploit military-to-commercial spin-offs in its communications, electronics, machine tool, and transportation sectors.[9] More recently, Scott Harold argued that the administration of President Lee Myung-bak (2008–13) sought to make "South Korea's defense industry into an 'engine of growth' that would average US$4 billion worth of exports per year and employ 50,000 people by 2020."[10]

Technonationalism and armaments production in the Asia-Pacific

More than any other factor, however, the technonationalist impulse appears to have driven defense industrialization in the Asia-Pacific. Technonationalism (a word first coined by Robert Reich in the 1980s)[11] is more than just a "security of supply" issue or a fancier word to describe protectionist economic and developmental policies. The technonationalist impulse is, of course, not limited to just armaments production or just to Asia. Many other industrial sectors in the region have benefitted from technonationalist policies, such as iron and steel, automobiles, electronics, shipbuilding, and the like. Japan and Korea, for example, have invested billions in building up domestic aircraft industries. Besides the Asia-Pacific, many nations around the world have pursued technonationalist policies and approaches when it comes to arms manufacturing. Beginning in the 1960s, Brazil embarked on an ambitious defense industrialization program, according to a national policy of *segurança e desenvolvimento* (security and development), which entailed large public investments in armaments as well as in other heavy industry and high-technology sectors.[12]

Nevertheless, there appears to be no good definition as to what constitutes "military technonationalism." At its most fundamental level, technonationalism simply entails the indigenous development of technology – as much for its own sake as for any economic benefits it might incur. As David Edgarton has put it, technonationalism was about countries, through indigenous technological development, trying to determine their place in the global pecking order, even if this was just "bragging rights."[13] At the heart of technonationalism is, of course, the nation-state:

> [N]ations are the units that innovate, that have R&D budgets and cultures of innovation, that diffuse and use technology. The success of nations, it is believed by techno-nationalists (who rarely if ever label themselves as such), is dependent on how well they do this.[14]

As military technonationalism has been defined by defense analysts and political economists such as Richard Samuels and Christopher Hughes, however, it has come to mean much more, at least in a military context.[15] In the particular case of armaments production, technonationalism is as much about securing geopolitical and strategic autonomy as it is about achieving technological and industrial self-sufficiency when it comes to defense. In other words, military technonationalism serves broad, bold national strategic ambitions, particularly the emergence of a country as a modern, independent, even powerful nation-state. Samuels argues that technonationalism is nothing less than the "struggle for independence and autonomy through the indigenization of technology."[16] It is, he adds, the "embrace of technology for national security."[17] Hughes describes technonationalism as "maximizing military technological autonomy in order to maximize national strategic autonomy."[18] Samm Tyroler-Cooper and Alison Peet, for their part, define the technonationalist model as "characterized by a focus on the development of indigenous capabilities for self-reliance and autonomy."[19] In short, technonationalism views autarky in military technology to be just as crucial to national security as is any particular weapon system.

Technonationalism in armaments production is particularly apropos for states aspiring to great power status. As Samuels notes, a nation-state cannot expect to be taken seriously unless it possesses a modern military, i.e. "rich nation/strong army." At the same time, an aspiring great power's armed forces may not be credible if it relies on other nations for the bulk of its weaponry. Therefore, to extend Samuel's "rich nation/strong army" analogy further, great nations have great arms industries. This line of reasoning has been particularly ubiquitous when it comes to Asian-Pacific armaments production: most large countries in the region – China, India, Japan, South Korea, Indonesia – have all attempted to create indigenous defense industries and to engage in ambitious arms-manufacturing programs in order to buttress their regional great-power ambitions.

Military technonationalism may have its roots in national security and economics, but it goes beyond that. It is about status – in this case, a particular nation's place in the international hierarchy of great powers. This is the appeal and power of military technonationalism, at least as it applies to indigenous armaments production: when the national security and economic arguments buttressing domestic weapons manufacturing fail, many nations still persist in pursuing autarky (and sometimes even "double down" in their commitments).

Technonationalism is more than an objective or a set of goals, however – it is also a plan of action. The technonationalist model contains its own strategy for achieving autarky in armaments production, one that, paradoxically, involves the exploitation of *imported* technologies in order to eventually realize self-sufficiency. This process usually entails the course of moving from *learning* to *innovating*, of going from *imitating* technology to *owning* and *advancing* technology – in this particular case, for the creation and promotion of a national indigenous defense industry. As the *Economist* puts it, "The focus is laid on national goals through accessing foreign technology and the monopolization of technology."[20]

Samuels divides the technonationalist process into three stages: indigenization, diffusion, and nurturing.[21] "Indigenization" refers to the acquisition of technology and its insertion into the local technological and industrial base; since this technology typically originates from foreign sources (e.g., through technology transfers or licensed-production), there is arguably a "technoglobalist" aspect to technonationalism at this phase, what some have described as a "techno-hybrid" model.[22] In any case, the technonationalist process is most critical for its "diffusion" and "nurturing" phases, in which the technology, however acquired, is assimilated and circulated throughout the national technology base and is further "processed" with localized inputs, i.e., indigenous R&D. The end result is that the technology has been changed and advanced sufficiently so that it is something new and innovative.

Japanese military technonationalism and the drive for *kokusanka*

Japan is perhaps the most closely associated with the technonationalist model when it comes to defense industrialization; consequently, it serves as an excellent starting point for any discussion and examination of military technonationalism in the Asia-Pacific. Richard Samuels' seminal work, *Rich Nation, Strong Army*, describes how central technonationalism was to the Japanese idea of *kokusanka*, or self-reliance, in arms manufacturing.[23] Wealth and power went hand in hand, and economic and technological development was crucial to the creation of a militarily strong state. For Japan, therefore, military technonationalism was the "embrace of technology for national security."[24] Industrial power and military strength were fused, and technology was the key. In subsequent studies of Japan's defense industry by Michael J. Green and Christopher Hughes, each also addresses the phenomenon of Japanese postwar armaments production through Samuels's technonationalist lens.[25]

No Asian nation is more synonymous with technonationalism than Japan, both in general terms of national economic development and specifically with regard to defense industrialization. Japan was the first country in Asia to industrialize and modernize, beginning in the late 19th century; during this initial phase technology imports and licensing were commonly used to shortcut the process of development. Imperial Japan, after the Meiji restoration, was also the first Asian state to pursue indigenous defense production in a major fashion. The new nationally established military – the Imperial Army and Navy – demanded the most sophisticated armaments available at time. Consequently, Tokyo pursued a decidedly technonationalist approach toward armaments production. It was during this period that it coined the slogan, "rich nation/strong army," capturing, as Samuels put, "the essence of military technonationalism."[26] As he further put it:

> Arms manufacturing, the most modern industrial sector before the Restoration, was the bellwether for Japan's forced march to industrialization. While light manufacturing prospered and provided foreign exchange, much of its machinery and production technology was developed in the military arsenals. The young Japanese oligarchy wanted to mobilize the resources of the economy to produce better weapons.[27]

Hughes echoes Samuel's argument, asserting that technonationalism and *kokusanka* goes back at least as far as the Meiji Restoration and establishment of modern Japan. The "ultimate objective" of a technonationalist arms-manufacturing strategy, he says, was always autarky.[28]

It was during the Imperial period (1868–1945) that the classic military technonationalist process of *nurturing, indigenization*, and *diffusion*, as laid out by Samuels, was first put into practice.[29] The new Imperial government promoted the development of a modern indigenous arms industry through high levels of defense spending. This not only benefitted government-run arsenals and shipyards, but it also encouraged private companies to enter into armaments manufacturing, although initially this centered mostly around supportive industries, such as iron and steel, automotives, fabrication, and the like. By the 1920s and 1930s, however, Japanese industry was producing a wide variety of military equipment, particularly such *zaibatsu* (vertical combines) as Mitsubishi and Kawasaki. In particular, commercial companies such as Nakajima, Mitsubishi, and Kawasaki came to dominate military aircraft development and manufacturing in Japan; this culminated in the production of some of the most advanced fighter aircraft during the World War II, particularly the A6M Zero and the B5N torpedo bomber. At the same time, the Japanese Imperial government worked to reduce its dependencies on foreign weapons systems, gradually replacing them with products of domestic design and manufacture. The Imperial Navy established the Navy Technology Department to identify and obtain foreign technologies that could be eventually indigenized; this included warships designs, propulsion systems, wireless communications, aircraft, and ordnance.[30] Japan also acquired licenses to manufacture foreign-designed military equipment, such as steam turbines (for warships), aircraft engines, and even entire systems (such as the L2D, a licensed-produce version of the Douglas DC-3), gradually replacing them with cloned, domestically produced versions. Links were also established between the military and private industry and universities to encourage diffusion and further indigenization of military know-how. Consequently, by the start of World War II Japan was more or less self-reliant in the design, development, and manufacture of armaments.

After its defeat in 1945, Japan was essentially demilitarized and disarmed, and much of its defense industry was converted to civilian production, such as motor scooters and railway carriages. With the outbreak of the Korean War (1950–53) and spread of the Cold War to Asia, however, US occupying forces permitted Japan to restart armaments production. According to Michael Green, US occupying forces spent more than 700 million yen rebuilding Japan's heavy industry, and the local arms industry quickly resumed the manufacture of munitions and other types of military equipment, the bulk of which was sold to the US military.[31] Key products included 81mm mortars, artillery shells, bombs and explosive powders, and spare parts for aircraft and vehicles, as well as aircraft repair and maintenance services for US armed forces.[32] Japan began to rearm in the early 1950s, and in 1954 the Self-Defense Force (SDF) was created. Around the same time *Keidanren* (the Japan Business Federation) established its Defense Production Committee (DPC), ostensibly to promote the development of local arms industries. However, it quickly came to exert considerable influence over SDF procurement and over Tokyo's policies and practices when it came to rearmament and indigenization. As Samuels put it: "The DPC wasted no time establishing itself as the political coordination and command center for Japanese defense and technology interests."[33]

This second postwar period of Japanese defense production has generally been characterized as the pursuit of *kokusanka*, or self-reliance, in arms manufacturing.[34] One cannot understand the motives behind post-1945 Japanese armaments production and defense industrialization without an appreciation for *kokusanka*, and one cannot understand *kokusanka* without an appreciation for the technonationalist impulse. As Green put it, "The drive for *kokusanka* can be explained first and foremost as a function of *technonationalism*."[35] To put it another way, in the case of postwar Japanese security policy, self-sufficiency and technonationalism are inseparable. Industrial power equals military power and therefore national security, and technology was the key. According to Samuels, "Japanese military and industrial strategies have been built on a fusion of industrial, technological, and national security policies,"[36] adding that military technonationalism was the "embrace of technology for national security."[37] Christopher Hughes echoes Samuels, arguing that autarky has always been the "ultimate objective" of a technonationalist arms-manufacturing strategy.[38]

In reconstituting its postwar defense industrial base, Japan pursued a classic incrementalist and globalist approach of moving from highly import-dependent types of armaments production to greater and greater autarky. The earliest produced armaments during this period were US weapons systems, built under license and often including, initially, at least, critical subsystems and components purchased directly from the United States. For example, Japan's first jet fighter, the F-86F Sabre, was licensed assembled by Mitsubishi Heavy Industries (MHI), with the engine and armaments, particularly the AIM-9B Sidewinder air-to-air missiles (AAM), acquired off the shelf from their US suppliers. This was followed by the full licensed-production of the F-104J, F-4EJ, and F-15J combat aircraft, including their jet engines and radar.[39] Other US weapons systems subsequently licensed produced in Japan included the AIM-7F and AIM-9L AAMs, CH-47 and UH-60/SH-60 helicopters, and the P-3 maritime patrol aircraft.

By the late 1950s, however, Japan was already attempting to replace or supplement the licensed-production of foreign-designed systems with indigenous products. The country designed and flew its first indigenous jet aircraft, the T-1 trainer, in 1958; altogether, 66 aircraft in this series were produced, and the last 20 were even powered by a domestically developed turbojet engine. The T-1 was followed by the T-2/F-1 supersonic trainer/close-support attack aircraft, of which a total of 167 were built in the early 1970s. The F-1 was a relatively simply ground attack and maritime interdiction aircraft, and 42 percent of its parts were produced under license, including its engine; nevertheless, it constituted Japan's first independently developed combat jet.[40] Similar indigenous development and production programs were undertaken in the missile sectors, e.g., the AAM-1, AAM-4, and AAM-5 air-to-air missiles; the Type-03 and Type-91 surface-to-air missiles; and the ASM-1 and ASM-3 anti-ship cruise missiles.

Japan's aircraft industry entered its next phase of *kokusanka* with the F-2 combat aircraft program, initiated in the 1980s as the FSX (Fighter Strike Experimental) project.[41] Originally, it was conceived as a true "Rising Sun" fighter, totally indigenous from stem to stern, optimized for air defense and maritime strike. It was supposed to incorporate the latest technology found in Japan's highly advanced industrial base, including the heavy use of nonmetal composites and a then-revolutionary active electronically scanned array (AESA) radar. However, US political pressure, together with Tokyo's growing realization that designing and developing a totally indigenous fighter would likely be extremely challenging (and likely prohibitively expensive), eventually forced Japan to abandoned this idea in favor of a much more modest codevelopment project with the United States. Consequently, in 1987 Tokyo and Washington signed a memorandum of understanding to codevelop the F-2, based on the already existing US F-16 Fighting Falcon. The F-2 bore a superficial resemblance to the F-16, but differed in many important aspects, including (1) a 25 percent larger wing area; (2) the use of composites (such carbon-fiber materials), to reduce the aircraft's weight and radar signature; (3) a larger tailplane and air intake; (4) a new canopy; and (5) a larger nose in order to accommodate the AESA radar.[42] In addition, MHI and Lockheed Martin, which produced the F-16, agreed to a 60:40 split in cost-sharing and work shares when it came to development and production. As such, Lockheed Martin was responsible for manufacturing the left-hand wingboxes as well as all aft fuselages and leading-edge wing flaps.[43] The F-2's first flight took place in 1995, and that same year Tokyo placed an order for 141 of these aircraft, to enter service starting in 1999.

Japanese efforts at *kokusanka* have perhaps been the most pronounced in the ordnance and shipbuilding sectors. Nearly every major weapon system in the Ground Self-Defense Force (GSDF) is of Japanese origin, from small arms to main battle tanks. Japan has been building its own tanks since the early 1960s. Its current main battle tank, the Type-10, is equipped with an indigenously developed 120mm smoothbore gun and modular ceramic composite armor. In addition, Japan has long produced its own destroyers, submarines, and amphibious ships. This includes the novel *Izumi*-class "helicopter destroyer," which is really a large helicopter-based amphibious assault vessel; the 19,000-ton *Izumi* features a through-deck design and below-deck hangars, and it closely resembles a small aircraft carrier, prompting speculation that it could be used as the basis for Japanese fixed-wing carrier, launching short-take and vertical landing (STOVL) combat aircraft like the F-35B Joint Strike Fighter (JSF). Japan also constructs its own submarines, the latest of which is the *Sōryū*-class. Armed with Harpoon anti-ship missiles and equipped with the Stirling engine for air-independent propulsion (which provides for longer periods of submerged sailing), the *Sōryū* is one of the most advanced conventionally powered submarines in the world.

Nearly unique to Asia, arms manufacturing in Japan has traditionally been pursued in a public-private partnership between the state and industry. As opposed to most other Asian-Pacific nations, which rely mostly on state-owned enterprises to manufacture armaments, nearly all defense production is embedded in a few large, highly diversified private companies, such as Mitsubishi Heavy Industries (MHI),

Mitsubishi Electric, Kawasaki Heavy Industries (KHI), etc. These companies in turn have traditionally been given a near-guarantee to some slice of the annual defense procurement budget, in which each company holds a monopoly stake in certain sectors; MHI, for example, is the country's only manufacturer of fighter jets, KHI is the sole builder of large airframes (such as the P-1 maritime patrol aircraft and the C-X transport plane), Fuji Heavy Industries produces only turboprop trainers, etc. In cases where there are two or more contractors – such as in shipbuilding (both MHI and KHI construct submarines, for example) – the firms tend to operate in a duopoly, alternating production contracts. In one case, two competitors – Ishikawa-jima Harima Heavy Industries (IHI) and Sumitomo Heavy Industries – merged their naval shipbuilding businesses into a single company, Marine United.

Moreover, Japanese military technonationalism has always stressed a high degree of civil-military integration. Both Samuels and Hughes note that arms production in Japan traditionally entailed a dual-use approach to technology: technology was consciously intended to "interdiffuse" between the military and civilian sectors.[44] According to Samuels, the Japanese process of defense innovation involved the deliberate spread of advanced technologies beyond their original function and intent, via "joint ventures, technology exchange agreements, cross-licensing, second-sourcing, production-sharing," etc.[45] As Hughes put it, "a key and constant feature of this drive for autonomous defense production has been to promote indigenous production . . . in tandem with integration where possible of civilian and military defense production."[46] Consequently, technological breakthroughs and progress in armaments production were *supposed* to spin off to commercial projects (such as military aircraft manufacturing helping to support firms in their efforts to expand into commercial aerospace, or through providing offshoot technologies that found their way into Japanese "bullet-train" programs); at the same time, commercial R&D was supposed to be spun on to military work (Samuels notes that advances in Japan's civilian microchip industry had direct military applications when it came to airborne active phased-array radar systems).[47]

All in all, Tokyo has put considerable resources into building up and maintaining a technologically advanced domestic arms industry, and the "indigenization" of defense production has long been national policy. On the surface, this practice has been highly successful: Japan is capable of building its own locally designed tanks, armored vehicles, warships, submarines, and various missile systems. Consequently, the Japan Self-Defense Force is almost entirely self-sufficient in military equipment. Where it has been forced to import, often because the cost of indigenous development was too high, Tokyo has secured licenses to manufacture these weapons in Japan; even then, the long-term goal has always been to eventually replace licensed-production with domestically developed systems.

Military technonationalism in other Asian-Pacific countries

Japan is perhaps the epitome of the technonationalist model of armaments production, and most other Asian-Pacific countries, wittingly or not, have largely copied the Japanese approach when it comes to defense industrialization and the indigenization of arms acquisitions.

China, for example, has long pursued a technonationalist strategy when it comes to indigenous arms production. Bates Gill and Taeho Kim observed that the Chinese quest for autarky in armaments goes back to the 19th century, noting that modernizers during that time knew that China could "not simply import complete weapons systems but also [had to] learn from foreign production techniques in order to establish a self-sufficiency in arms production."[48] This became the basis for the so-called *tiyong* concept of the Qing dynasty: *zhongxue weiti, xixue weiyong* ("Chinese learning for substance, Western learning for use"); inherent in this concept is the idea that China should exploit foreign technology as much as it can in order to wean itself eventually off of it.[49]

The idea of being autonomous in the development and manufacture of arms has been no less critical under the present-day People's Republic of China (PRC).[50] Self-reliance (*zili gengsheng*) remains even more so an "indispensable component . . . of national security,"[51] and as such the PRC's defense industry has always been "geared toward the objective of autonomy."[52] Even though the PRC in its early years (1949–1961) had to rely heavily upon the Soviet Union for military technology (both weapons systems and production capacities), it was always China's long-term goal to "return to the first path" of self-reliance.[53] If anything, China's unhappy experiences with foreign military-technical assistance – i.e., the abrupt cutoff of Soviet military aid in the early 1960s and the Western arms embargo following the 1989 Tiananmen Square crackdown – only reinforced its natural impulses to become eventually self-reliant in arms production.[54] By the early 1970s, therefore, the Chinese were actively engaged in the development of a number of indigenously designed weapon systems, including fighter aircraft, ballistic and anti-ship cruise missiles, tanks, surface combatants, and submarines (and, of course, nuclear weapons).

The current phrase to express this desire for autarky in defense production and acquisition is, according to Tai Ming Cheung, *zizhu chuangxin*, or "innovation with Chinese characteristics" (also sometimes translated as "indigenous innovation," "autonomous innovation," and "self-reliant innovation").[55] According to Cheung, *zizhu chuangxin* is a "core aspiration" of China's political, military, and defense-industrial leadership, and, in particular, it has been formalized in the 2006–2020 Medium- and Long-Term Defense Science and Technology Development Plan.[56] It basically entails four broad approaches toward technological innovation and development: (1) introduce, (2) digest, (3) assimilate, and (4) re-innovate (what Cheung dubs the IDAR strategy). Interestingly, China's IDAR strategy very closely resembles the same model that Samuels uses to describe the Japanese technonationalist guidelines of indigenization-diffusion-nurturing.[57]

Self-reliance has long been a fundamental goal of indigenous armaments production in **India**. As Ajay Singh put it:

> After independence, and the adoption of a policy of non-alignment, it was . . . obvious that foreign policy would need to be reinforced by a policy of self-reliance in defense . . . Prime Minister Jawaharlal Nehru

believed that no country was truly independent, unless it was independent in matters of armaments.[58]

When it comes to indigenous armaments production, India makes an important distinction between "self-sufficiency" and "self-reliance." *Self-sufficiency* is defined as those "stages in defense production (starting from design to manufacture, including raw materials) . . . be carried out within the country." *Self-reliance* refers simply to the indigenous armaments production, and it can include the continued import of foreign designs, technologies, systems, and manufacturing know-how.[59] Of the two, the former was the more desirable from an economic, military, and especially political perspective, since it anticipated a higher degree of independence and autonomy. While self-sufficiency was the preferred end state, however, India has long had to accept a high level of "self-reliance" in indigenous defense manufacturing as a military and industrial expediency. Consequently, large amounts of foreign military technology had to be imported in order to create and nurture an Indian military-industrial complex.[60]

Nevertheless, it has always been India's policy to replace self-reliance with true self-sufficiency[61] As far back as the 1950s India began to build its own indigenous systems, including a fighter jet, i.e., the HF-24 *Marut*. However, truly indigenous armaments production did not really take off until the 1980s, with the inauguration of several ambitious homegrown projects, such as the *Tejas* Light Combat Aircraft, the *Arjun* tank, and, especially, the multi-pronged Integrated Guided Missile Development Program (IGMDP). While many of these so-called indigenous programs still incorporated considerable amounts of foreign technology or subsystems, the objective has always been to gradually but determinedly reduce this dependency and eventually to achieve "self-sufficiency."[62] This intent is evident in several initiatives launched over the past 20 years to increase the "local content" found in locally produced weapons systems, albeit with limited success so far.[63]

Technonationalism is also evident in the **Republic of Korea's** historical approach to indigenous armaments production. The ROK is committed to a strategy of "cooperative self-reliant defense,"[64] including the goal of "acquiring the ability to independently develop primary weapon systems for core force capability."[65] Moreover, the South Korean government sees an advanced domestic defense industry as an important symbol of the country's "coming of age," both as a high-technology powerhouse and as a regional power. Consequently, Seoul places a strong emphasis on a "domestic weapons first" policy, a course of action that goes back to the early 1970s and the implementation of the Yulgok Project, an ambitious program of defense industrialization that was intended to lay down "a basic foundation for a self-defense capability for the 21st century."[66]

South Korea arms manufacturing began in the early 1970s, with the assembly of M16 assault rifles under license from the United States. Local armaments production expanded greatly after the promulgation of the Nixon Doctrine, which reduced US defense commitments to Asia but at the same time liberalized the export of advanced military technologies to Asian allies. Consequently, Seoul invested billions of dollars into the domestic development and production of fixed wing and rotary aircraft, missile systems, tanks and armored vehicles, artillery systems, large surface warships, and submarines. As a result, whereas in 1985 only 59 percent of South Korea's arms were procured domestically, by 1995 this amount has risen to nearly 80 percent.[67]

The ROK has particularly promoted its aerospace sector as a key strategic industry. It has poured billions of dollars into indigenous combat aircraft programs, starting with the licensed-production of F-5 and later F-16 fighters, and the MD-500 light helicopter. In the late 1980s Korea began developing a light turboprop trainer plane, the KT-1, followed by the T-50 supersonic jet trainer/light attack aircraft. More recently, the ROK has begun work on an advanced fighter jet, the so-called KFX program, which is intended to enter service sometime during the 2020s. These projects, along with other military and commercial programs (including a planned 90-seat passenger plane), are part of a long-term strategy to propel the South Korean aircraft industry into the world's top 15 global aerospace producers by 2020.

Seoul also expects to greatly expand its defense exports, an area where it has traditionally been a minor player. South Korea has, for example, sold tanks and artillery systems to Turkey and infantry fighting vehicles to Malaysia. It has also enjoyed some success marketing its military aircraft; it has sold KT-1 trainer planes to Indonesia, Turkey, and Peru, and T-50 jets to Indonesia, Iraq, and the Philippines.

Indonesia undertook armaments production in earnest in the mid-1970s, with the establishment of several state-owned "strategic enterprises," the most important of which were PT *Industri Pewsawat Terban Nusantara* or IPTN (aircraft), PT PAL (shipbuilding) and PT Pindad (small arms and munitions).[68] Under Suharto, Jakarta viewed armaments production both as a way to overcome the country's backward state of industrial and technological development and as a means to leapfrog the country into the forefront of regional great powers. Suharto was particularly influenced by his minister for research and technology, B.J. Habibie (who later succeeded Suharto as president of the republic). Habibie, an aerospace engineer by training, explicitly viewed the establishment of an aerospace industry as both an instrument and a model for advancing the country's overall technology and industrial base.[69] For him, IPTN in particular was to serve as an indicator of Indonesia's intentions to become a modern industrialized nation and "to prove that a Third World, Muslim-majority country could make a hi-tech leap into global aviation."[70] Just as important, a powerful defense industry was intended to make Indonesia into a military power to be reckoned with in Southeast Asia. IPTN began by license-assembling helicopters and light transport planes, and later manufacturing components for F-16 fighters and British Hawk trainers were acquired by the Indonesia air force. In the early 1980s it entered into a joint venture with CASA of Spain to codevelop and manufacture the CN-235 military/commercial transport aircraft (which was subsequently also exported to the United Arab Emirates, Brunei, Malaysia, Pakistan, South Korea, and Thailand). In addition, PT PAL built patrol boats for the Indonesian navy, while PT Pindad manufactured assault rifles and pistols and developed the *Panser*, an indigenous 6x6 wheeled armored personnel.

Taiwan has long espoused the need for a "self-reliant national defense," and therefore the need for a strong domestic arms industry.[71] It was "important to develop a self-defense capability without relying on outside assistance."[72] Consequently, this stress on a self-reliant defense translated into an ambitious program of indigenous military R&D and production dating back to the 1970s, aimed at meeting specific national defense requirements; in the mid-1990s, for example, a Taiwanese defense minister argued that self-sufficiency was Taipei's "ultimate goal" in developing an indigenous defense industry, adding that the country would never import weaponry that could be developed or manufactured locally.[73] In 1999, then-President Lee Teng-hui asserted that while Taiwan may acquire some arms from abroad, "only by enhancing self-development capability can national security be ensured."[74] In 2000 the former head of the Chung Shan Institute of Science and Technology (CSIST) declared that, rather than what kinds of weapons Taiwan imports, "what mainland China really cares [about] is which new weapons and new defense technology [Taiwan] can develop." CSIST, he therefore asserted, "has become one of our bargaining chips in negotiating with the mainland."[75]

This policy of maintaining a strong domestic defense industrial base, capable of delivering to the Taiwanese military state-of-the-art weaponry in critical key sectors (such as missile systems), has continued well into the 21st century.

Finally, even smaller countries like **Malaysia** and **Singapore** have promoted the idea of producing indigenous weapons systems, even if in limited scope and numbers. Singapore in particular has regarded a robust indigenous defense industry to be an important compensation for its small size and vulnerability ("its midget psychosis," so to speak[76]) with regard to its larger neighbors.

Technonationalism and the state

Not surprisingly, technonationalism in general, and military technonationalism in particular, greatly stresses the role of the state and downplays market forces – in this case, when it comes to cultivating local arms industries. Governments are usually intimately and actively involved in the technonationalist process of defense industrialization.[77] In the Asia-Pacific, consequently, the state has typically played an instrumental role in the establishment and nurturing of indigenous arms industries. In many instances, armaments production has been either wholly or partly dominated by the state through military-run or state-owned and -operated enterprises, such as are found in China, India, and Indonesia. However, even when weapons manufacturing is embedded in private industry – such as in Japan or South Korea – the government generally underwrites armaments production via direct investments, tax incentives, monopoly sourcing, guaranteed military contracts, and the like.[78] In the case of Japan, for example, the government encouraged local industries to participate in arms manufacturing by selecting sole-source suppliers, ensuring production contracts, guaranteeing (low but steady-state) profits, providing *de facto* R&D subsidies, and permitting opportunities for spin-off into more lucrative commercial endeavors (such as commercial aerospace).[79] In South Korea the government encouraged firms to enter into arms production through a variety of incentives – such as tax breaks, low-interest loans, and direct subsidies – and coercive measures – such as tying defense contracting to state support for engaging in other types of commercial production.[80] In addition, the Korean government, through the state-owned Korea Finance Corporation, controls a 26.4 percent stake in Korea Aerospace Industries (KAI), the country's leading military aircraft manufacturer.

In addition, the state has typically assumed most, if not all, of the risk and costs for weapons development and production by intervening on the supply side to mandate and fund indigenous solutions for a military hardware requirement. Throughout most of the Asia-Pacific region, for example, state-run defense R&D institutes undertake the actual design and development of military systems that are subsequently provided to arms factories for production. This strategy is particularly pronounced in the case of regional aerospace industries, where the bulk of production has traditionally been directed toward meeting domestic defense demands.

Technonationalism and the role of foreign technology

When countries – and especially emerging arms producers – decide to enter into indigenous arms manufacturing, they have tended to follow roughly similar patterns of industrialization and production. This process usually entails a series of gradual and progressive steps leading to greater sophistication and self-sufficiency in the design, development, and manufacturing of weapon systems. As such, it has often been described as the "ladder of production," and while scholars may disagree as to how many steps there are in this process or the precise ordering of these stages of production, the idea that countries engage in an evolutionary and incremental mode of defense industrialization is broadly accepted.[81]

According to the ladder of production, indigenous arms production is a process of transitioning from extremely high to very low levels of foreign dependency for weapons and production technologies. Initial armaments production tends to rely heavily on imported technical assistance from countries possessing already well-advanced defense industries. Most second-tier arms-producing countries start out by assembling weapon systems from imported parts and components (knock-down kits). The next step usually consists of the licensed-production of foreign weapon systems, with some (and, in many cases, eventually nearly all) of the actual manufacturing of components and subsystems performed indigenously. This is usually followed by limited indigenous development and production of relatively simple, "low-tech" armaments – such as small arms, ordnance, or small patrol boats – along with the codevelopment of more sophisticated armaments in partnership with more advanced foreign arms producers. Particularly at these later stages in the ladder of production model, basic arms-manufacturing capabilities are increasingly supplemented by incremental improvements in the country's independent military R&D base. Accordingly, a country may attempt to indigenously produce more complex (i.e., "mid-tech") weapon systems, such as light armored vehicles or trainer aircraft. Lastly, a country may attempt to design and develop its own advanced weapon systems – such as fighter aircraft, missiles, submarines, large surface combatants, or military electronics – either across the board or by carving out certain niches or specialties.

There is no disingenuousness in the relationship between the military technonationalist model and the used of foreign technology in order to advance the cause of autarky. Indeed, most technonationalists freely concede the necessity of importing technology – considerable amounts of it, in fact, especially at the beginning – and, historically, most cases of technonationalist-based defense industrialization have depended quite heavily on foreign technology and know-how. As Samuels put it:

> Self-reliance in technology does not mean autarky . . . [it] refers to the ability to absorb all needed technologies, and the attainment of this self-reliance is attained not at a stroke but in stages.[82]

Therefore, technonationalism and technoglobalism are not necessarily incompatible with each other; indeed, they complement each other.[83] *All* aspiring arms-producing nations in the Asia-Pacific have used foreign technologies to some extent in order to learn, leapfrog, reduce costs, shorten R&D timeframes, and avoid technological blind alleys. Post-war Japan and South Korea, for example, relied heavily upon the United States for defense technology and production know-how in their initial arms-production phases, particularly when it came to the licensed-production of US military systems. India depended on considerable aid and assistance from the Soviet Union, Britain, and France when it began the process of indigenous defense industrialization.

The PRC's initial arms-manufacturing base was essentially a copy of the Soviet model, entailing not only the licensed-production of Soviet weaponry but doing so in turnkey factories provided wholesale by the USSR. Even as late as the 1980s and 1990s, Beijing frequently resorted to foreign suppliers for technologies that it could exploit for national military-industrial development. During the 1980s, for example, China imported helicopters from France, missiles from Italy, and radar systems from the United Kingdom; many of these systems were subsequently reverse engineered and produced in Chinese factories. The United States had a joint development program ("Peace Pearl") with China to upgrade the latter's J-8 fighter jet, and Washington also provided China with torpedoes and artillery-locating radar that were likely exploited later on for technological gain.[84] After the 1989 arms embargo (imposed on China by the United States and the European Union in response to the Tiananmen Square massacre), China turned then to Russia, and during the 1990s and early 2000s it imported considerable amounts of defense technology from Russia. Beijing acquired a license to locally produce the Su-27 fighter jet, which it later copied and now illegally manufactures as the J-11B. China has also exploited Russian defense technologies in such areas as air-to-air, air-to-ground, and surface-to-air missiles, precision-guided munitions, submarines, and even manned space systems.[85]

At the same time, the ultimate goal was always autarky in innovation. Foreign dependencies were generally perceived to be tactical and short-term, a "necessary evil" in order to underwrite indigenization and development (Samuels, for example, asserts that the Japanese saw licensed-production as a "middle road" toward autarky).[86] Military technonationalism (like technonationalism in general), therefore, does not mean total autarky, at least, not initially: it permits the acquisition of foreign technologies should they provide positive benefits (i.e., shortcuts to technological advancement). As Samuels put it in the case of Japan:

> International collaboration and indigenization do not work at cross-purposes in Japanese practice . . . Indeed, a key lesson of the Japanese case is that national systems of innovation can be integrated in the global economy without sacrificing their integrity. Japanese scientific and technological networks have long been linked internationally in ways that actually reinforce nationalism. International networks neither obviate the relevance of national systems [of innovation] nor dilute technonational incentives. . .[G]lobalization will not obviate national systems of innovation.[87]

In short, technoglobalism is acceptable and permissible if it leads to the eventual objective of self-sufficiency.

The appeal and challenges of military technonationalism

As we will see in the subsequent chapters, the technonationalist model has a strong appeal for states wishing to enter into defense production. Becoming self-reliant in armaments is a tempting goal, and as such many countries in the Asia-Pacific have pursued the Japanese model toward autarky. Even in the 21st century the "siren song" of military technonationalism is still very powerful, as we shall see in the following chapters examining the process of indigenous arms manufacturing in several newly emerging industrialized states in the Asia-Pacific. It is important to note, too, that technonationalism and the quest for autarky in armaments is not limited to just states in the Asia-Pacific; a diverse collection of countries around the world – including Turkey, Iran, Pakistan, and Brazil – also harbor ambitious goals to create robust, technologically sophisticated indigenous defense industries.

At the same time, we shall also see that military technonationalism has not been without its challenges. The application of technonationalism to armaments production in the Asia-Pacific has experienced many setbacks and failures, not the least of which was due to overconfidence and subsequent overreach. Autarky in armaments is an extremely elusive thing, and enthusiasm and throwing money at a problem is no shortcut to success. In general, therefore, it is permissible to question whether the technonationalist approach is a wise strategy for defense industrialization – if not in every case, then at least in situations where a nation's aspirations far outstrip its capacities to commit sufficient resources and to create large enough markets.

Notes

1 For a fuller discussion of the various tiers of arms-producing nations, see Richard A. Bitzinger, *New Ways of Thinking about the Global Arms Industry: Dealing with "Limited Autarky"* (Canberra: Australian Strategic Policy Institute, November 2015), pp. 2–4.
2 First-tier producer-states consist of critical defense-technological innovators, like the United States, and sophisticated "adaptors and modifiers," like the larger European arms producers (e.g., Britain, France, Germany, and perhaps Italy). Keith Krause, *Arms and The State: Patterns of Military Production and Trade* (Cambridge: Cambridge University Press, 1992), pp. 26–33; Andrew L. Ross, "Full Circle: Conventional Proliferation, the International Arms Trade and Third World Arms Exports," in Kwang-il Baek, Ronald. D. McLaurin, and Chung-in Moon, eds., *The Dilemma of Third World Defense Industries* (Boulder, CO: Westview Press, 1989), pp. 1–31; Richard A. Bitzinger, *Towards a Brave New Arms Industry?* (Oxford: Oxford University Press, 2003), pp. 6–7.
3 Carol Evans, "Reappraising Third-World Arms Production," *Survival*, March/April 1986, pp. 100–101; Janne E. Nolan, *Military Industry in Taiwan and South Korea* (New York: St. Martin's Press, 1986), pp. 12–14; Ralph Sanders, *Arms Industries: New Suppliers and Regional Security* (Washington, DC: NDU Press, 1990), pp. 11–17; Michael Brzoska and Thomas Ohlson, "Conclusions," in Michael Brzoska and Thomas Ohlson, eds., *Arms Production in the Third World 1971–1985* (Oxford: Oxford University Press, 1987), pp. 279–280.
4 Susan Willett, "East Asia's Changing Defense Industry," *Survival*, Autumn 1997, p. 114; Tim Huxley and Susan Willett, *Arming East Asia* (Oxford: Oxford University Press, July 1999), p. 51.
5 Jong Chul Choi, "South Korea," in Ravinder Pal Singh, ed., *Arms Procurement Decision Making, Volume I: China, India, Israel, Japan, South Korea and Thailand* (Oxford: Oxford University Press, 1998); Patrice Franko-Jones, *The Brazilian Defense Industry* (Boulder, CO: Westview Press, 1992), pp. 55–63; Janne E. Nolan, "South Korea: Ambitious Client of the United States," in Michael Brzoska and Thomas Ohlson, eds., *Arms Production in the Third World 1971–1985* (Oxford: Oxford University Press, 1987), pp. 218–219.
6 Nolan, "South Korea: Ambitious Client of the United States," pp. 218–219; Choi, "South Korea," p. 191.
7 James Elliot and Ezio Bonsignore, "Asia's 'New' Aerospace Industry: At the Turning Point?" *Military Technology*, February 1998, p. 31; Dean Cheng and Michael W. Chinworth, "The Teeth of the Little Tigers: Offsets, Defense Production and Economic Development in South Korea and Taiwan," in Stephen Martin, ed., *The Economics of Offsets: Defense Procurement and Countertrade* (London: Harwood, 1996), pp. 245–246.
8 Steinberg, "Israel: High-Technology Roulette," pp. 168–170; Sadeh, "The Israeli Defense Industry: The End Is Nigh?" p. 3.
9 ROK Ministry of National Defense, *Defense White Paper 1993–1994* (Seoul: Korea Institute for Defense Analyses, 1994), p. 192.
10 "Global Insider: Defense Exports an Essential Part of 'Global Korea: An Interview with Scott W. Harold," *World Politics Review*, October 21, 2013 (http://www.worldpoliticsreview.com/articles/print/13318).

11 Robert Reich, "The Rise of Technonationalism." *Atlantic Monthly*, May 1987.
12 Patrice Franko-Jones, *The Brazilian Defense Industry* (Boulder, CO: Westview Press, 1992), pp. 55–63.
13 David E.H. Edgerton, "The Contradictions of Techno-Nationalism and Techno-Globalism: A Historical Perspective," *New Global Studies*, Vol. 1, No. 1 (2007), pp. 2–4.
14 Edgerton, "The Contradictions of Techno-Nationalism and Techno-Globalism," p. 1.
15 Richard J. Samuels, *Rich Nation, Strong Army: National Security and the Technological Transformation of Japan* (Ithaca, NY: Cornell University Press, 1994); Christopher W. Hughes, "The Slow Death of Japanese Techno-Nationalism?" *Journal of Strategic Studies*, Vol. 34, No. 3 (June 2011).
16 Samuels, *Rich Nation, Strong Army*, p. ix.
17 Samuels, *Rich Nation, Strong Army*, p. 31.
18 Hughes, "The Slow Death of Japanese Techno-Nationalism," p. 453.
19 Samm Tyroler-Cooper and Alison Peet, "The Chinese Aviation Industry: Techno-Hybrid Patterns of Development in the C919 Program," *Journal of Strategic Studies*, Vol. 34, No. 3 (June 2011), p. 385.
20 "Techno-Nationalism," *Figuring Things Out*, December 14, 2011 (http://dinakarr. blogspot.sg/2011/12/techno-nationalism.html).
21 Samuels, *Rich Nation, Strong Army*, pp. 33, 42–56.
22 Tyroler-Cooper and Peet, "The Chinese Aviation Industry," pp. 385–387.
23 Richard J. Samuels, *Rich Nation, Strong Army: National Security and the Technological Transformation of Japan* (Ithaca, NY: Cornell University Press, 1994).
24 Samuels, *Rich Nation, Strong Army*, p. 31.
25 Michael J. Green, *Arming Japan: Defense Production, Alliance Politics, and the Postwar Searth for Autonomy* (New York: Columbia University Press, 1995); Christopher W. Hughes, "The Slow Death of Japanese Techno-Nationalism?" *Journal of Strategic Studies*, Vol. 34, No. 3 (June 2011). See also Michael E. Chinworth, *Inside Japan's Defense: Technology, Economics, and Strategy* (Washington, DC: Brasseys, 1992).
26 Samuels, *Rich Nation, Strong Army*, p. 84.
27 Samuels, *Rich Nation, Strong Army*, p. 84.
28 Hughes: "The Slow Death of Japanese Techno-Nationalism?" pp. 454.
29 Samuels, *Rich Nation, Strong Army*, pp. 85–93.
30 Samuels, *Rich Nation, Strong Army*, p. 91.
31 Michael J. Green, *Arming Japan: Defense Production, Alliance Politics, and the Postwar Search for Autonomy* (New York: Columbia University Press, 1995), pp. 32–33.
32 Samuels, *Rich Nation, Strong Army*, pp. 138–141.
33 Samuels, *Rich Nation, Strong Army*, p. 144.
34 The best, most recent studies of Japanese postwar defense industries and armaments production include Samuels, *Rich Nation, Strong Army*; Green, *Arming Japan;* Michael E. Chinworth, *Inside Japan's Defense: Technology, Economics, and Strategy* (Washington, DC: Brasseys, 1992); and Christopher W. Hughes, "The Slow Death of Japanese Techno-Nationalism? Emerging Comparative Lessons for China's Defense Production," *Journal of Strategic Studies*, Vol. 31, No. 3 (July 2011), pp. 451–477.
35 Green, *Arming Japan*, p. 11.
36 Richard J. Samuels, "Reinventing Security: Japan Since Meiji," in Edward R. Beauchamp, ed., *Japan's Role in International Politics since World War II* (New York: Routledge, 2011), p. 47.
37 Samuels, *Rich Nation, Strong Army*, p. 31.

38 Hughes, "The Slow Death of Japanese Techno-Nationalism?" p. 454.
39 Arthur Alexander, *Of Tanks and Toyotas: An Assessment of Japan's Defense Industry* (Santa Monica, CA: RAND, 1993), pp. 38–42.
40 Bill Gunston and Mike Spick, *Modern Combat Aircraft* (New York: Crescent Books, 1983), pp. 136–137; Alexander, *Of Tanks and Toyotas*, p. 39.
41 For an excellent analysis of the F-2 program, see Mark Lorell, *Troubled Partnership: A History of US-Japanese Collaboration on the FS-X Fighter* (Santa Monica, CA: RAND, 1995).
42 Jenny Lu, "Technology Transfer and the F-2 Fighter: How the Japanese Defense Industry Defied the Odds," *Pointer*, Vol. 38, No. 4 (2012), pp. 58–60.
43 Tom Breen, "Lockheed Martin Starts Beefing up Work Force for Japan's F-2," *Defense Daily*, October 21, 1996.
44 Samuels, *Rich Nation, Strong Army*, p. 49.
45 Samuels, *Rich Nation, Strong Army*, p. 51.
46 Hughes, "The Slow Death of Japanese Techno-Nationalism?" p. 453.
47 Samuels, *Rich Nation, Strong Army*, pp. 288–301.
48 Bates Gill and Taeho Kim, *China's Arms Acquisitions from Abroad: A Quest for "Superb and Secret Weapons"* (Stockholm: Stockholm International Peace Research Institute, 1995), p. 11.
49 Gill and Kim, *China's Arms Acquisitions from Abroad*, p. 2.
50 See Richard A. Bitzinger, et al., "Locating China's Place in the Global Defense Economy," in Tai Ming Cheung, ed., *Forging China's Military Might: A New Framework for Assessing Innovation* (Baltimore, MD: Johns Hopkins University Press, 2014); Tai Ming Cheung, "The Chinese Defense Economy's Long March from Imitation to Innovation," *Journal of Strategic Studies*, Vol. 34, No. 3 (June 2011); James Mulvenon and Rebecca Samm Tyroler-Cooper, *China's Defense Industry on the Path of Reform*, prepared for the US-China Economic and Security Review Commission, October 2009; Tai Ming Cheung, "Dragon on the Horizon: China's Defense Industrial Renaissance," *Journal of Strategic Studies*, Vol. 32, No. 1 (February 2009); Tai Ming Cheung, *Fortifying China: The Struggle to Build a Modern Defense Economy* (Ithaca, NY: Cornell University Press, 2008); Evan S. Medeiros, et al., *A New Direction for China's Defense Industry* (Santa Monica, CA: RAND, 2005); Keith Crane, et al., *Modernizing China's Military* (Santa Monica, CA: RAND, 2005).
51 Han S. Park and Kyung A. Park, "Ideology and Security: Self-Reliance in China and North Korea," in Edward E. Azar and Chung-in Moon, eds., *National Security in the Third World: The Management of Internal and External Threats* (Aldershot: Edward Elgar, 1988), p. 119.
52 J.D. Kenneth Boutin, "Arms and Autonomy: The Limits to China's Defense-Industrial Transformation," in Richard A. Bitzinger, ed., *The Modern Defense Industry* (Santa Barbara, CA: ABC-CLIO, 2009), p. 213.
53 Crane, et al., *Modernizing China's Military*, p. 137.
54 Gill and Kim, *China's Arms Acquisitions from Abroad*, pp. 8–47.
55 Cheung, "The Chinese Defense Economy's Long March," p. 326.
56 Cheung, "The Chinese Defense Economy's Long March," pp. 326–327.
57 Tai Ming Cheung, *The Role of Foreign Technology Transfers in China's Defense Research, Development, and Acquisition Process*, Policy Brief 2014–5, January 2014, Institute on Global Conflict and Cooperation, San Diego, CA, 2014, pp. 2–3.
58 Ajay Singh, "Quest for Self-Reliance," in Jasjit Singh, ed., *India's Defense Spending* (New Delhi: Knowledge World, 2000), pp. 126–127.
59 Singh, "Quest for Self-Reliance," p. 127.
60 Angathevar Baskaran, "The Role of Offsets in Indian Defense Procurement Policy," in J. Brauer and J.P. Dunne, eds., *Arms Trade and Economic Development:*

Theory, Policy, and Cases in Arms Trade Offsets (London: Routledge, 2004), pp. 211–213, 221–226.

61 Manjeet S. Pardesi and Ron Matthews, "India's Tortuous Road to Defense-Industrial Self-Reliance," *Defense & Security Analysis*, Vol. 23, No. 4 (December 2007), pp. 421–429; Ron Matthews and Alma Lozano, "India's Defense Acquisition and Offsets Policy," in Rajesh Basrur, ed., *India's Military Modernization: Strategic Technologies and Weapons Systems* (Oxford: Oxford University Press, 2014).

62 Richard A. Bitzinger, *Towards a Brave New Arms Industry?*, pp. 16–18.

63 Singh, "Quest for Self-Reliance," p. 151.

64 Noh, Hoon, *South Korea's "Cooperative Self-Reliant Defense": Goals and Directions*, KIDA Paper No. 10 (Seoul: Korea Institute for Defense Analyses, April 2005), p. 5.

65 ROK Ministry of National Defense, *Defense White Paper 1999: Republic of Korea* (Seoul: Korea Institute for Defense Analyses, 1999), p. 145.

66 Choi, "South Korea," p. 183.

67 Choi, "South Korea," p. 185.

68 *IHS Jane's Navigating the Emerging Markets: Indonesia* (London: IHS Global Ltd, January 2012), pp. 20–23; Bilveer Singh, "ASEAN's Arms Industries: Potential and Limits", *Comparative Strategy*, Vol. 8, No. 2 (1989), p. 251.

69 John Bailey, "Habibie's Grand Design," *Flight International*, February 19, 1992, pp. 51–52.

70 Margot Cohen, "New Flight Plan," *Far Eastern Economic Review*, March 2, 2000, p. 45.

71 A.J. Gregor, R.E. Harkavy, and S.G. Neuman, "Taiwan: Dependent 'Self-Reliance,'" in Michael Brzoska and Thomas Ohlson, eds., *Arms Production in the Third World* (London: Taylor & Francis, 1986), pp. 233–49.

72 Sun Wen, "Towards a Modernized National Defense," in Jason C. Hu, ed., *Quiet Revolutions in Taiwan* (Taipei: Kwang Hwa Publishing, 1994), p. 428.

73 *CNA* (Taipei), April 17, 1996.

74 *AFP* (Hong Kong), July 1, 1999.

75 "Former CSIST Chief Calls for Domestic Military," *CNA* (internet version), in English, August 8, 2000.

76 Singh, "ASEAN's Arms Industries," p. 254.

77 Edgerton, "The Contradictions of Techno-Nationalism and Techno-Globalism," p. 1.

78 John Grevalt, "Seoul Aborts Sales of Korea Aerospace Industries," *Jane's Defense Weekly*, September 3, 2012.

79 Hughes, "The Slow Death of Japanese Techno-Nationalism?" pp. 457–458; Green, *Arming Japan*, pp. 17–18.

80 Kwang-il Baek and Chung-in Moon, "Technological Dependence, Supplier Control and Strategies for Recipient Autonomy: The Case of South Korea," in Kwang-il Baek, Ronald. D. McLaurin, and Chung-in Moon, eds., *The Dilemma of Third World Defense Industries* (Boulder, CO: Westview Press, 1989), pp. 158–159.

81 Krause, *Arms and the State*, p. 171; Michael Brzoska and Thomas Ohlson, "Arms Production in the Third World: An Overview," in Michael Brzoska and Thomas Ohlson, eds., *Arms Production in the Third World 1971–1985* (Oxford: Oxford University Press, 1987), pp. 15–27; James Everett Katz, "Understanding Arms Production in Developing Countries," in James Everett Katz, ed., *Arms Production in Developing Countries: An Analysis of Decision Making* (Lexington, MA: Lexington Books, 1984), pp. 8–9; Willett, "East Asia's Changing Defense Industry," pp. 116–118.

82 Samuels, *Rich Nation, Strong Army*, p. 45.

83 Tyroler-Cooper and Peet, "The Chinese Aviation Industry," pp. 385–387.

84 United States General Accounting Office, *China: U.S. and European Union Arms Sales Since the 1989 Embargoes*, GAO/T-NSIAD-98–171 (Washington: US General Accounting Office, April 28, 1998).

85 Richard Fisher, Jr., *PLAAF Equipment Trends*, International Assessment and Strategy Center, October 30, 2001 (http://www.strategycenter.net/research/pubID.8/pub_detail.asp).

86 Samuels, *Rich Nation, Strong Army*, p. 46.

87 Samuels, *Rich Nation, Strong Army*, p. 340.

2 Military modernization in the Asia-Pacific

An overview

The Asia-Pacific is a leading consumer of arms. Increasingly, some of the most modern and most advanced armaments are finding their way into the inventories of Asian militaries. As a result, many Asian-Pacific militaries have experienced a significant, if not unprecedented, build-up over past several years, both in terms of quantity and quality. In an arc stretching from Japan to India, regional armed forces have been adding new capabilities for war-fighting, and therefore the capacities for new military roles and missions. Local navies have acquired advanced warships, operating both on the surface and under the seas, that provide the potential for force projection and expeditionary operations. Regional air forces have purchased modern combat aircraft, many armed for the first time with weaponry providing the capacity to engage in air-to-air combat beyond visual range, as well as undertaking precision-guided air-to-ground missions. Ground forces, meanwhile, have acquired new weapons systems, ordnance, and military equipment that greatly increase their firepower, lethality, and mobility. Finally, regional militaries are gaining both the hardware and the software needed to improve their capacities for surveillance, reconnaissance, target acquisition, and command and control.

This arms-acquisition process has been impelled by a number of strategic and economic factors. The drive for great power status, whether regionally and globally, has pushed many Asian-Pacific countries to build up their militaries as a hard-power complement to their growing "soft-power" (particularly economic) capacities. These developments have in turn sparked a competition in arming and counter-arming – even drawing in those countries that seek only to acquire improved defenses against increasingly assertive, well-armed neighbors – that some have dubbed an arms race. Regional great power machinations have been further complicated by the United States' renewed interest in the Asia-Pacific, as evidenced in Washington's "pivot to Asia" and its subsequent growing military presence. At the same time, rising regional defense budgets, driven by growing economies, together with a "buyers' market" in the global arms marketplace offering almost every type of advanced weaponry, have made it possible for most nations in the Asia-Pacific to acquire at least some number of modern armaments. Finally – and this key development will be the focus of the remainder of this book – these countries are increasingly supplementing or *replacing* foreign-sourced weaponry with homegrown systems that are, in many instances, as good as imported arms. This combination of strategic competitions, rising regional wealth, and growing availability of advanced conventional weaponry have created a "harmonic convergence" underwriting one of the most far-reaching arms build-up in the world.

The regional political-military context behind military modernization

Countries in the Asia-Pacific have many reasons for acquiring new defense hardware and improving national military capabilities. The region is clearly one of constantly shifting security dynamics, with rising great powers (China and India), new threats and security challenges (missile attacks, terrorism, the proliferation of weapons of mass destruction (WMD), international crime, and the like), ongoing territorial disputes (for example, in the South China Sea, around the Senkaku/Diaoyu and the Dokdo/Takashima Islands, and the unresolved China-Taiwan dispute), and new military commitments (such as disaster relief, humanitarian assistance, and contingency and stabilization operations) that require new capabilities for power projection, mobility, firepower, intelligence and surveillance, and joint operations. All of these factors in one way or another are affecting regional military modernization activities.

China in particular possesses great power aspirations that drive much of its requirements for a modern military, particularly when it comes to projecting sustained power beyond its border, delivering firepower and precision-strike, and dominating the information battlespace. Beijing, for example, seeks to gain "hard" power commensurate with growing "soft" power (i.e., economic, diplomatic, and cultural). Naturally, China still seeks sufficient military capabilities to deal with any "Taiwan contingency" – that is, the ability to isolate the island, to invade, and to occupy it if necessary – and to engage in anti-access/area denial operations to interdict US forces seeking to come to the island's defense. Additionally, China wants to be able to press its territorial and exclusive economic zone (EEZ) claims in the East and South China seas. Finally, Beijing increasingly seeks military means to police and protect sea lines of communication (SLOCs) in order to safeguard Chinese shipping and trade as well as to secure energy supplies, given China's reliance on oil and gas imports.

These goals are clearly apparent in China's increasingly assertive, even belligerent, behavior in the South China Sea. Beijing is actively engaged in significantly militarizing the South China Sea, including aggressive patrolling by naval and para-naval forces; the dramatic expansion of military defenses (i.e., long-range surface-to-air missiles) on Woody Island, China's largest possession in the South China Sea; and, in particular, an ambitious artificial island-building program that has taken place in the Spratlys over the last few years, including constructing runways on at least three reefs, emplacing radar stations, and even temporarily moving weapons to these islands.

At the same time, China is keen to build expeditionary forces capable of projecting power out to the "second island chain," which is delineated by Guam,

Indonesia, and Australia. Eventually it hopes to be able to project sustainable force throughout the whole of the Western Pacific and into the Indian Ocean. In particular, this goal has led Beijing to de-emphasize ground forces in favor of building up the naval, air, and missile forces of the People's Liberation Army (PLA). According to its 2015 white paper, the PLA will continue to de-emphasize land operations, all but abandoning People's War (except in name and in terms of political propaganda), particularly in favor of giving new stress and importance to sea power and force projection: "The traditional mentality that land outweighs sea must be abandoned, and great importance has to be attached to managing the seas and oceans and protecting maritime rights and interests."[1] As a result, the PLA Navy (PLAN) "will gradually shift its focus from 'offshore waters defense' to the combination of 'offshore waters defense' with 'open seas protection,'"[2] an evolutionary development from what was announced in the 2006 white paper, which proclaimed that the "Navy aims at gradual extension of the strategic depth for offshore defensive operations."[3] This will require a "combined, multi-functional and efficient marine combat force structure. The PLAN will enhance its capabilities for strategic deterrence and counterattack, maritime maneuvers, joint operations at sea, comprehensive defense and comprehensive support."[4]

The 2015 white paper also stated that PLA Air Force (PLAAF) would "endeavor to shift its focus from territorial air defense to both defense and offense, and build an air-space defense force structure that can meet the requirements of informationized operations."[5] This included building up the PLAAF's capacities for strategic early warning, air-carried precision-strike, air and missile defense, "information countermeasures," and strategic force projection (i.e., airlift).

As initially laid out in its 2004 defense white paper, the PLA officially embraced the doctrine of fighting and winning "limited local wars under conditions of 'informatization.'"[6] Under such conditions, such warfare was still seen – as with earlier concepts of "limited local war" – as revolving around short-duration, high-intensity conflicts characterized by agility, speed, and long-range attack. As a further refinement, though, "limited local wars under conditions of 'informatization'" entailed joint operations fought simultaneously throughout the entire air, land, sea, space, and cyber battlespace (i.e., five-dimensional warfare), and relying heavily upon extremely lethal high-technology weapons. Such an operational doctrine also emphasizes preemption, surprise, and "shock value," given that the earliest stages of conflict may be crucial to the outcome of a war.[7] Consequently, "limited local wars under conditions of 'informatization'" stressed mobility, flexibility, power projection, precision-strike, and joint operations fought throughout the entire battlespace.[8]

China's evolved thinking about "informationized warfare" was further elaborated in the 2015 white paper, which places even greater emphasis on both cyber operations and space war. Cyberspace, it argued, "has become a new pillar of economic and social development, and a new domain of national security . . . As cyberspace weighs more in military security, China will expedite the development of a cyber force, and enhance its capabilities of cyberspace situation awareness [and] cyber defense."[9] In addition, outer space "has become a commanding height in international strategic competition." Consequently, China plans to "keep abreast of the dynamics of outer space, deal with security threats and challenges in that domain, and secure its space assets to serve its national economic and social development, and maintain outer space security."[10]

China's military rise has helped sparked a growing Sino-American competition in the far western Pacific Ocean. At the beginning of 2012, the Obama administration formally promulgated its new "pivot," or rebalancing, back to the Asia-Pacific region. The pivot implies a consequential, realignment of US global power, emphasizing air- and sea-based operations in an "arc extending from the Western Pacific and East Asia into the Indian Ocean region and South Asia." In particular, this rebalancing involves the redeployment of US forces from other parts of the world. The US Navy (USN) plans to position 60 percent of its fleet in the Pacific Ocean compared to a current 50:50 split between the Pacific and Atlantic. In addition, 2,500 US Marines are to be based in Darwin, Australia, while Singapore has agreed to "host" up to four of the new USN Littoral Combat Ships. Finally, the United States is seeking to expand its access to ports and other facilities in the Philippines and Vietnam.

As part of the pivot, in September 2009 the US Navy and Air Force signed a classified memo to initiate an inter-service effort to develop a new joint operational concept, dubbed AirSea Battle (ASB). ASB was intended to preserve stability and to sustain US power projection and freedom of action, and to offset current and anticipated asymmetric threats through a novel integration of US Air Force and Navy's concepts, assets, and capabilities. Although ASB was redesignated as the "Joint Concept for Access and Maneuver in the Global Commons" (JAM-GC) in January 2015, it remains, for all practical extent and purposes, the same. To counter a hypothetical crisis scenario or conflict in which an adversary employs an A2/AD strategy, ASB/JAM-GC envisions a preemptive, stand-off, precision-strike – or "Networked, Integrated Attack-in-Depth" – initiated and carried out by US forces, by striking the enemy's intelligence, surveillance, and reconnaissance (ISR) assets from afar and thus denying them situational awareness; by suppressing and disrupting the enemy's air-defense networks using stealthy long-range platforms; and by conducting diverse follow-on operations, such as "distant blockades," in order to seize the operational initiative and to ensure protracted US freedom of action in the region.

With regards to the Asia-Pacific, ASB/JAM-GC would probably be used mostly to deal with the anti-access/area denial (A2/AD) conundrum posed by China. In fact, ASB/JAM-GC is, at least implicitly, first and foremost about countering China's supposedly growing abilities to "no-go" sanctuaries in the far western Pacific Ocean, particularly in and around the East and South China Seas. Consequently, ASB/JAM-GC, together with Washington's increasing emphasis on FONOPS (freedom of navigation operations) in the South China Sea, is a critical indicator of a growing strategic competition between the United States and China.

Great-power ambitions and rivalries can be found in other Asian-Pacific nations. India, too, is striving for regional status, particularly in terms of asserting military strength throughout the Indian Ocean area. In particular, this entails the protection of local SLOCs (90 percent of the country's trade, and most notably most of its oil and gas supplies, transits through the Indian Ocean), sovereignty enforcement, and sea area denial to its adversaries.[11] India is increasingly keen to build national capacities for power projection, long-range surveillance and intelligence, and expeditionary warfare.[12] These great-power aspirations are manifested in the country's efforts to develop its nascent nuclear forces (including land- and sea-based missile delivery systems). New Delhi is increasingly concerned as to the growing Chinese presence in the Indian Ocean, while at the same time being aware of its increasing growing "post-modern" security interests, such as humanitarian assistance, disaster relief, and securing the "global commons." As Geoffrey Till has put it: "Over the past few years, there has been a notable expansion of India's internationalist concerns. This is reflected in its growing role in globalization."[13]

Many of these taskings particularly favor the Indian Navy, which in turn is increasingly becoming the high-tech focus of the military. Consequently, the Indian Navy has increasingly attached greater stress to force projection capabilities for the purposes of "sea-based deterrence, economic and energy security, forward presence and naval diplomacy."[14] Nevertheless, according to Till there still exists within the Indian military (particularly the navy) a very traditional, "modernist" impulse, emphasizing such things as "sea control," which in turn "confers independence of action."[15] Till also claims that:

> The two impulses are not necessarily incompatible with each other, however: [The Indian Navy's] current transformational emphasis is on developing their power projection and expeditionary capabilities, much of which could be deployed in the general defense of the system alongside their maritime partners.[16]

Japan's current military modernization efforts are driven by three factors: the need to deal with new emerging regional threats, the requirement for increased interoperability with an expeditionary US military, and the desire for the "normalization" of Japanese foreign and defense policy. As the threat of a Soviet attack on the Japanese mainland has disappeared, other security concerns have arisen, particularly missile threats from North Korea, international terrorism, and instability in regions far from Japan that could nevertheless affect Japanese economic, political, and military security. At the same time, within Northeast Asia China has become a growing security concern, as evidenced by large Chinese military exercises that took place in April 2010 near Japanese territorial waters,[17] as well as by the ongoing dispute over the Senkaku/Diaoyutai Islands. Additionally, Japan is a critical US ally in Asia and, as such, serves as a forward operating area for US forces in the region. The Japan Self-Defense Forces (SDF) are also increasingly partners with US forces (and by extension, with NATO and Australian forces) in contingency operations, such as security-building in the Indian Ocean, patrolling in the Straits of Malacca, and stabilization operations in Iraq and Afghanistan.

Consequently, Japan's security interests have expanded far beyond Northeast Asia, and the SDF have accordingly greatly increased their mobility and expeditionary capabilities, firepower, and systems for command, control, communication, computer, intelligence, surveillance, and reconnaissance (C4ISR).

Japan is also keen to pursue a foreign and defense policy more befitting a "normal" nation, and Tokyo has permitted the country's military to play a larger and more active role in regional security missions. Japan's conservative political parties, the preeminent Liberal Democratic Party (LDP) in particular, have long sought to upgrade the status of the SDF and to legitimize its role as an outright military force. In addition, many in the LDP and elsewhere have called for the revision of Japan's so-called Peace Constitution in order to explicitly permit the maintenance of self-defense forces and to allow these forces to be used in international peacekeeping and security operations.

Military modernization activities within the smaller countries of the Asia-Pacific have enjoyed their own particular momentum. The Republic of Korea, in addition to being confronted by a possibly growing threat from North Korea, also has growing pretensions of becoming a regional power. As such, Seoul is pursuing military acquisition programs intended to increase the capacities of the ROK Armed Forces in order to permit them to act more independently of the US military and in support of a more nationalistic, self-reliant, and self-assertive foreign and defense policy.[18] This is apparent in South Korea's efforts to acquire a blue water navy (complete with a large fleet of ocean-going submarines) that will rival Japan's and China's maritime forces.

In Southeast Asia, there is growing unease over China's aforementioned "creeping assertiveness" in the South China Sea and its growing military presence in the region.[19] Additionally, Southeast Asian countries face new unconventional threats, particularly piracy, terrorism, international crime, and human trafficking. At the same time, many Southeast Asian states are just as often suspicious of each other as they are of external powers such as China, with historical animosities still existing between Malaysia and Singapore, Malaysia and Indonesia, and Thailand and Burma, to name but a few. Moreover, competing claims over EEZs in the South China Sea and over the Spratly Islands are just as strong between the various Southeast Asian nations as they are between these nations and Beijing. Consequently, these tensions have also been powerful motivators behind recent national military build-ups in the region, especially when it comes to acquiring capabilities – particularly long-range naval and air forces – for patrolling and protecting EEZs and promoting sovereignty rights.[20]

Defense spending in Asia-Pacific

Certainly, rising military budgets have underwritten a significant regional arms build-up. Asian defense expenditures have increased significantly since the turn of the century. China, for example, has experienced real (i.e., after inflation) double-digit increases in defense expenditures nearly every year since 1997. Between 1997 and 2016, real Chinese military expenditures grew, on average, around 10 percent per annum. By 2016, China's defense budget totaled US$146 billion, making China the second-highest military spender in the world, outstripping Japan, France, Russia, and the United Kingdom when it comes to national defense. Overall, Chinese military expenditures have grown more than 600 percent in real terms since the late 1990s, permitting Beijing to put considerable resources into the hardware and software of military modernization.

Other Asia-Pacific nations have similarly increased defense expenditures over the past decade and a half. Indian defense spending, for example, has grown by 87 percent between 2000 and 2015, according to data provided by the Stockholm International Peace Research Institute (SIPRI).[21] In 2015, Indian military expenditures totaled approximately US$51.3 billion.[22] South Korea's defense budget has grown by 65 percent, reaching US$38.6 billion in 2015. Malaysia's military budget (according to SIPRI data) reached US$4.9 billion, more than doubling since 2000. Indonesian defense spending over the same period grew from US$1.1 billion to US$7.6 billion, at least a four-fold increase. For its part, Singapore's defense expenditures has also more than doubled, from US$4.3 billion in 2000 to US$5.8 billion in 2008 (again, in constant 2005 dollars); in 2010, Singapore's military budget totaled US$9.7 billion. Altogether, military spending in Southeast Asia rose by at least 50 percent in real terms since the turn of the century.[23] Of all the other larger countries in the Asia-Pacific, only Japan and Taiwan have needed to contend with relatively static military budgets. Even so, Taipei was recently able to secure additional funding for arms acquisitions, and in 2008 it finally closed a US$6.5 billion deal with the United States for the purchase of several types of weapons systems, including Patriot PAC-3 missiles, AH-64 attack helicopters, Harpoon anti-ship cruise missiles (ASCMs), and Javelin antitank guided weapons. This deal also included a major upgrade of F-5 and F-16 fighter jets in the Taiwanese air force. In 2010 Taiwan made a follow-on, US$6 billion buy of additional missiles, helicopters, and further F-16 upgrades, and in 2014, Washington agreed to sell Taipei four additional *Perry*-class frigates.

Quantifying military modernization in Asia

Over the past two decades or so, many countries throughout the Asia-Pacific have initiated ambitious military modernization programs (Table 2.1). As a result of both indigenous production and arms imports, these militaries have gained or are currently gaining capabilities that they did not possess earlier, such as new means for power projection, precision-strike, long-range attack, lethality and firepower, and, in particular, battlespace intelligence, communications, and command and control. These new capabilities are especially due to recent acquisitions by Asian countries of modern surface combatants, amphibious assault vessels, aircraft carriers, submarines, advanced fighter aircraft armed with long-range air-to-air missiles, missile defenses, and a host of new precision-guided munitions, including anti-ship and land-attack cruise missiles, stand-off weapons, and smart bombs. Additionally, these weapons acquisitions are being complemented by greatly improved C4ISR systems, including unmanned aerial vehicles and drones, airborne early warning aircraft, and state-of-the-art communications networks.

Table 2.1 Recent and planned major Asian-Pacific arms acquisitions

Country	Surface Combatants	Amphibious Ships/ Aircraft Carriers	Submarines	Combat Aircraft	Missiles & Other Systems
China	22+ Type-051C/-052B/-052C/-052D destroyers 4 Russian-built Sovremennyy-class destroyers 26+ Type-054/-054A frigates 60+ Houbei-class FAC(M)	1 Liaoning-class (ex-Varyag) Will likely build indigenous a/c carriers 4+ Type-071 LPDs LHD-class vessel reportedly under construction	26+ Song-/Yuan-class submarines (some w/ AIP) 12 Russian-built Kilo-class submarines 4+ Type-093 SSN 4 Type-094 SSBN	~300 Su-27/-30 fighters (some Su-27s locally produced) Building 400+ J-10 fighters J-31/J-35 5th-generation fighters under development	AAM: R-77, PL-12 ASCM: 3M-54E/E1 Sunburn, 3M-80E Moskit, YJ-83 LACM: DH-10 SSMs: DF-11/-15
India	3 Kolkata-class destroyers Building 4 Visakhapatnam-class destroyers Plans to build 7 Project-17A-class frigates	Acquiring ex-Russian Kiev-class STOVL aircraft carrier, to be modified to fly MiG-29 fighters Building Indigenous Aircraft Carrier, INS Vikrant, to fly MiG-29 or Tejas fighters	Acquiring 6+ French-designed Scorpène-class submarines; later submarines AIP-equipped 3 Arihant-class nuclear-powered submarines under construction	Acquiring 240+ Su-30MKI fighters (some locally produced) 36 Rafale fighters Building up to 260 locally produced Tejas fighters	AAM: R-77 ASCM: Exocet, Brahmos SSMs: Prithvi, Agni
Indonesia	Acquiring 2+ Sigma-class frigates 4 Dutch-built Sigma-class corvettes	Acquiring 4 Korean-made LDPs	3 Korean-built Type-209 submarines	16 Su-27/-30 fighters 24 ex-USAF F-16s	AAM: R-77 ASCM: YJ-83
Japan	4 Akizuki-class destroyers 6 Kongo- and Atago-class destroyers, equipped with upgraded Aegis combat system and SM-3 missile for MD	3 Osumi-class LPDs 2 Hyuga-class DDH (14,000 ton); could be upgraded to LHD 2 Izumi-class (19,500-ton) DDH under construction (fixed-wing capable?)	Building 22 Soryu-class submarines (w/AIP)	94 F-2 fighter jets 42 F-35 JSF on order Indigenous 5th-gen fighter under development	AAM: AMRAAM, AAM-5 ASCM: Harpoon AGM: JDAM

Country	Surface ships	Submarines	Aircraft	Missiles
South Korea	Building 6 KDX-III destroyers, equipped with Aegis combat system, SM-2 air-defense missile; could be upgraded to MD capability; 3 KDX-I and 6 KDX-II destroyers; Building 2 *Dokdo*-class LPDs	9 German-designed Type-209 submarines, acquired 1990s; Building 9 German-designed Type-214 submarines (w/AIP)	61 F-15K fighters; 160 F-16 fighters; 40 F-35 JSF on order; Indigenous 4th+ gen fighter under development	AAM: AMRAAM; ASCM: Harpoon, *Haesung*; LACM: *Hyunmoo*-IIIC; AGM: JDAM, JASSM
Malaysia	2 British-built *Lekiu*-class frigates; 6 German-designed, locally built MEKO A100 OPVs; Acquiring 6+ French *Gowind*-class corvettes	2 French-built *Scorpène*-class submarines	18 Su-30MKM fighter; Plans to acquire 18 additional fighters, type undecided	AAMs: R-77; ASCM: Exocet; MRL: ASTROS-II
Philippines	Acquiring 3 ex-USCG cutters (re-classed as a frigate); 10 ex-Japanese Coast Guard ships		Acquiring 12 Korean-built FA-50 light combat aircraft	
Singapore	6 French-designed *Formidable*-class "stealth" frigates; 8 1,200-ton littoral combat ships under construction; 4 *Endurance*-class LPDs; Joint Multi-Mission Ship on order	4 ex-Swedish A-12 submarines; 2 ex-Swedish A-17 submarines (w/AIP); 2 German Type-218S submarines (w/AIP) on order	24 F-15S fighters; 74 F-16 Block 52/52+ fighters; Partner in Joint Strike Fighter (F-35) program, may acquire up to 100 F-35s	AAMs: AMRAAM, Python IV, AIM-9X; ASCM: Harpoon; AGM: JSOW, JDAM; MRL: HIMARS
Taiwan	4 ex-*Kidd*-class destroyers; 8 *Perry*-class frigates; 6 *Lafayette*-class frigates; 4 ex-*Knox*-class frigates, acquired 2000s; Building 30 *Kuang Hua* VI-class FAC(M); Developing *Hsun Hai* corvette; 1 ex-*Anchorage*-class LSD	Requirement for up to 8 submarines, but acquisition uncertain	150 F-16A/B fighters (being upgraded); 60 *Mirage*-2000 fighters; 130 locally built *Ching-kuo* fighters (being upgraded)	AAM: AMRAAM, AIM-9M, MICA, Magic II, Sky Sword I/II; AGM: Maverick; ASCM: Harpoon, *Hsiung Feng* II/III; MD: PAC-2/-3, Skybow III; LACM: *Hsiung Feng*-IIe

(*Continued*)

Table 2.1 (Continued)

Country	Surface Combatants	Amphibious Ships/ Aircraft Carriers	Submarines	Combat Aircraft	Missiles & Other Systems
Thailand	2 Chinese-built Type-053 frigates	1 Spanish-built STOVL aircraft carrier, equipped with AV-8A STOVL fighters (inoperable)	Requirement for 2+ submarines	12 Gripen fighters	AAM: AMRAAM
Vietnam	2 Russian-built *Gepard*-class frigates		Acquiring 6 Kilo-class submarines, with LACM	12 Su-27 fighters 36 Su-30MK2V fighters	AAM: R-77 ASCM: Kh-35/SS-N-25 Switchblade

Source: Compiled by author

Glossary
AAM: air-to-air missile
AGM: air-to-ground munition
AIP: air-independent propulsion
ASCM: anti-ship cruise missile
DDH: helicopter destroyer
FAC(M): fast-attack craft (missile-carrying)
LACM: land-attack cruise missile
LHD: landing helicopter dock
LPD: land platform dock
MD: missile defense
MRL: multiple-rocket launcher
OPV: offshore patrol vessel
SSM: surface-to-surface missile
SSN: nuclear-powered attack submarine
SSBN: nuclear-powered ballistic-missile submarine
STOVL: short takeoff/vertical landing
USCG: US Coast Guard

Sea power developments

Asia-Pacific navies have expanded considerably over the past 10 to 15 years, both in terms of quantity and, more importantly, in terms of capabilities. Many navies in the region that were once oriented mainly toward coastal defense – the so-called brown waters – are being upgraded to green water or even blue water (open ocean) capacities. Many countries in the region have consequently added larger surface combatants to their fleets, which greatly extends their range of operations as well as their sustainability and firepower. At the same time, many local navies either have expanded or are in the process of expanding their capacities for force projection and expeditionary warfare, in particular via the acquisition of platforms capable of operating rotary-wing and, increasingly, fixed-wing aircraft.

In the late 1990s, the Chinese People's Liberation Army Navy acquired four *Sovremennyy*-class destroyers from Russia. These ships are outfitted with the 3M-80E *Moskit* (also known as SS-N-22 Sunburn) ramjet-powered, supersonic, anti-ship cruise missile (ASCM), which has a range of 120 kilometers. Newer missiles have a 200-kilometer range. Since the turn of the century, however, the PLAN has increasingly relied upon its indigenous shipyards to supply it with modern warships. Since 2000 China has constructed as least 22 modern destroyers of the Type-51 and Type-052 class. The most important of these are the Type-052C and Type-052D, which are outfitted with *Aegis*-type air-defense radar and fire-control systems, as well as HHQ-9 surface-to-air missiles (SAMs), housed in vertical launch systems (VLS). These destroyers are also equipped with the indigenous YJ-83 or YJ-62 anti-ship cruise missile and the HN-2 land-attack cruise missile (a variant of the Russian Kh-55 missile). China has also added more than two dozen new frigates to its forces – particularly the Type-054A *Jiangkai*-class, which features a stealthy design and is armed with ASCMs and VLS-deployed SAMs – as well as the new-generation Type-022 *Houbei*-class catamaran-hulled missile fast attack craft (outfitted with YJ-83 ASCMs), of which at least 60 have been built.

In perhaps its most dramatic development, the PLAN has recently taken delivery of China's first aircraft carrier: the former, rebuilt Soviet carrier *Varyag*. A casualty of the post-Cold War, the *Varyag* was laid down in the early 1980s, but construction was halted in 1992 when the vessel was only 70 percent complete. Ukraine, which inherited it after the breakup of the Soviet Union, stripped the ship bare and left it exposed to the elements for several years. When the *Varyag* was finally sold and delivered to China in 2001 – ostensibly to be turned into a Macau casino – it was a rusted shell, without engines, rudder, weapons systems, or electronics. In addition, the process of removing sensitive equipment from the vessel had resulted in damage to its structure, so that even its seaworthiness was questioned. In mid-2005, however, the Chinese moved the *Varyag* to a dry dock at the Dalian shipyard in northeast China, where it underwent substantial repairs and reconstruction, along with the installation of new engines, radars, and electrical systems. The rebuilt ex-*Varyag* carrier underwent its first sea trials under PLAN colors in August 2011, and it was subsequently commissioned the *Liaoning* and accepted into service with the PLAN in 2012. The *Liaoning* is equipped with the J-15 fixed-wing fighter jet (reportedly reverse engineered from a Su-33 acquired surreptitiously from Ukraine), along with anti-submarine warfare and airborne early-warning helicopters.

The *Liaoning* will likely be used more as a research and training platform for future Chinese carrier designs and crews rather than as a fully functioning carrier (although the carrier could be pressed into military service in a limited capacity). At the same time, China is expected to build several indigenous carriers – as of 2016, in fact, at least one (and possibly two) homegrown carriers are under construction – and it is likely that the PLAN could acquire up to six aircraft carriers. If and when that happens, it would likely mean the reorientation of the PLAN around Carrier Battle Groups (CVBGs), with the carrier at the heart of a constellation of supporting submarines, destroyers, and frigates – an amalgamation of power projection at its foremost. Such CVBGs are among the most impressive instruments of military power, in terms of sustained, far-reaching, and expeditionary offensive force, and such a development would constitute a major shift in PLAN strategic direction.

China is also in the process of expanding its capacities for force projection and expeditionary warfare, in particular involving the acquisition of platforms capable of operating fixed-wing aircraft. China has launched at least four Type-071 17,000- to 20,000-ton LPD (landing platform dock) amphibious warfare ships, equipped with two helicopters and two air-cushioned landing craft (LCAC), and capable of carrying up to 800 troops; up to eight Type-071s could eventually be built. A larger LHD-type (landing helicopter dock) amphibious assault ship is also speculated.

The Japan Maritime Self-Defense Force (MSDF) has also been expanding its capacities for power projection through the acquisition of high-speed sealift ships (for logistics and transport) and three large amphibious *Osumi*-class ships. Ostensibly designated as an LST (landing ship tank), the *Osumi*-class vessel is of a size and design more resembling a LPD (including a large open deck for helicopters). The 13,000-ton *Osumi* can carry 330 troops and up to 10 tanks, and is outfitted with 4 helicopters and 2 LCAC hovercraft transports. Additionally, the MSDF is currently acquiring four so-called "helicopter destroyers" (DDH), ostensibly for anti-submarine warfare. In the last 2000s, the MSDF took delivery of two 19,000-ton *Hyuga*-class DDHs; these ships were subsequently supplemented with two larger (27,000-ton) *Izumi*-class DDHs, constructed in the early 2010s. Both of these classes of ship feature a through-deck design and below-deck hangars and thus resemble small aircraft carriers; currently, however, the *Hyuga* and *Izumi* are intended only for use with helicopters.

The Republic of Korea Navy (ROKN) has contracted for three 7700-ton KDX-III destroyers (also known as the *King Sejong the Great*–class), and it recently exercised an option for three additional vessels. The KDX-III is a vast improvement over other destroyers in the ROKN, being equipped with the U.S.-supplied Aegis air-defense radar and fire-control system as well as the Standard SM-2 Block IIIB air-defense missile. The KDX-III could be upgraded to the SM-3 missile for anti-tactical ballistic missile operations, although this is currently not being planned. In addition, the KDX-III is armed with the *Hyunmoo*-IIIC land-attack cruise missile (LACM) and either the Harpoon or the indigenous *Haesung* (Sea Star) ASCM. All these missiles are housed in 128 vertical launch cells.

In addition, South Korea is in the process of accepting into service the *Dokdo*-class amphibious assault vessel. The *Dokdo*-class LPX (landing platform experimental) displaces 14,000 tons and is capable of carrying 700 troops, 10 tanks, 15 helicopters, and 2 LCACs.[24] The *Dokdo* is intended to serve as a multifunctional vessel, in particular serving as a fleet command ship for a "rapid response fleet" that would include one *Dokdo*-class ship and several destroyers, frigates, and submarines.[25] Two *Dokdo*-class vessels have been ordered – designed and built by Korean shipyards – and the first was commissioned into the ROKN in 2007; plans to acquire a third ship in this class were subsequently cancelled, however.

While Japan and South Korea have no plans at the moment to acquire fixed-wing aircraft carriers, their current classes of open flight-deck helicopter ships – the *Izumi* and the *Dokdo* – could conceivably be modified (for example, by adding a "ski-jump" deck) to serve as pocket carriers capable of operating short takeoff/vertical landing (STOVL) combat jets, such as the F-35B Joint Strike Fighter (JSF).

For its part, India is already putting considerable resources into building a carrier-centered navy and is currently in the process of replacing its two aging British-built carriers. In the first place, the navy is acquiring the Soviet-built *Admiral Gorshkov*, a 45,000-ton *Kiev*-class carrier decommissioned by the Russian Navy in 1996. After several years of strenuous negotiations, Moscow and New Delhi agreed to a deal whereby Russia would provide the carrier *gratis*, while India would pay the Russia approximately US$1 billion to refit and upgrade the vessel to be capable of flying navy MiG-29 fighters off its deck in a STOBAR (short takeoff but assisted recovery) configuration.[26] This entailed stripping off the weaponry from the ship's foredeck and adding a 14.3 degree ski-jump on the bow and three arrestor wires on the angled landing deck. In addition, India would pay another US$700 million toward the aircraft and weapons systems, which include 12 single-seat MiG-29K Fulcrum-D fighter jets, four dual-seat MiG-29KUB trainer aircraft, and 6 Kamov Ka-27 and Ka-31 helicopters, along with training, simulators, spare parts, and maintenance facilities.[27] The carrier, renamed the INS *Vikramaditya*, was supposed to have been delivered to the Indian Navy in mid-2008, but refitting the vessel has turned out to be much more challenging than originally envisioned, resulting in considerable cost overruns – Moscow has asked for an additional US$1.2 billion to finish the upgrade – and delays. Eventually, however, the *Vikramaditya* was commissioned in late 2013 and entered into service in 2014.

The Indian Navy experienced similar teething problems with its indigenous aircraft carrier (IAC), formerly known as the air defense ship (ADS). The IAC, designated the INS *Vikrant*, is a 37,500-ton vessel and will utilize a STOBAR arrangement of ski-jump and arrestor wires and operate either the MiG-29K or India's indigenous *Tejas* Light Combat Aircraft (LCA), currently in development. Construction began in 2005 at the Cochin shipyards, but production problems have delayed the first IAC's in-service date by several years, which is now not expected until 2018, at the earliest.[28] Ultimately, however, the navy wants to operate a two- to three-carrier battle group force – one on each coast – with one in reserve.[29]

The move from brown water to open-ocean navies has been particularly pronounced in Southeast Asia. In the early 2000s, the Republic of Singapore Navy (RSN) acquired six *Formidable*-class frigates (based on the French *Lafayette*-class stealth vessel), armed with Harpoon ASCMs and the French Aster-15 air-defense missile; these frigates are a significant increase in the RSN's power-projection capabilities. Singapore is also currently building eight 1,200-ton littoral combat vessels. Indonesia is currently acquiring four new *Sigma*-class corvettes from the Netherlands, equipped with Chinese C-802 (YJ-83) ASCMs, while Malaysia is presently building six French-designed *Gowind*-class littoral combat ships; previously acquisitions include six MEKO-A100-class offshore patrol vessels as well as two British-built *Lekiu*-class frigates.

Several Southeast Asian nations have also been acquiring ships for expeditionary and amphibious warfare. The RSN, for example, operates four indigenously designed and constructed *Endurance*-class landing ships, each capable of carrying 350 troops, 18 tanks, 4 helicopters, and 4 landing craft; these ships will be supplemented by a new Joint Multi-Mission Ship, basically a helicopter assault ship (LHD). Indonesia (as well as the Philippines) has bought LPDs built in South Korean shipyards. Finally, it is worth noting that the Royal Thai Navy is the only other navy in Asia besides China and India to operate a fixed-wing aircraft carrier, the 10,000-ton, Spanish-built HTMS *Chakri Nareubet*. This vessel was originally outfitted with nine used AV-8A Harrier STOVL jets and six S-70B Seahawk helicopters. The *Chakri Nareubet* is rarely put to sea, however, due to its high operating costs, and the ship's Harrier jets were retired around 2008, leaving it basically functioning as a helicopter carrier. Thailand has also acquired one *Endurance*-class landing ship from Singapore.

The procurement of submarines is another area where Asia-Pacific nations have invested considerable time and effort. In some cases, nations that never before had submarines forces are now acquiring their first boats. China has greatly expanded its submarine fleet over the past 15 years. In the late 1990s and early 2000s, the Chinese navy acquired 12 *Kilo*-class diesel-electric submarines from Russia, some of which are armed with the 3M-54E *Klub* (SS-N-27) ASCM and the 53–65KE wake-homing torpedo. In addition, starting in the late 1990s the PLAN has acquired at least 26 Type-039 *Song*-class and Type-41 *Yuan*-class diesel-electric submarines.

Additionally, the PLAN has begun to replace its small and aging fleet of nuclear-powered submarines, i.e., five *Han*-class nuclear-powered attack boats (SSN) and one *Xia*-class nuclear-powered ballistic missile-carrying submarine (SSBN). The first in a new class of SSNs, the Type-093 *Shang*-class, was launched in 2002 and commissioned in 2006; at least four additional Type-093 are believed to have since entered service. The PLAN has also launched at least four new SSBNs of the Type-094 *Jin*-class, each carrying 12 JL-2 submarine-launched ballistic missiles (SLBMs) with a range of 7,000 kilometers (three times greater than that of the

JL-1 SLBM carried by the *Xia*). Follow-on classes of both new nuclear-powered submarines are expected over the next decade.

Japan is currently building a new class of diesel-electric submarines (the *Sōryū*) equipped with the Swedish-designed Stirling engine system for air-independent propulsion (AIP). At least eight *Sōryū*-class submarines are under construction and four more are planned (for a total of twelve boats), to be built at a rate of approximately one submarine a year.

South Korea is also increasing its submarine fleet. During the 1990s the ROKN acquired nine German-designed Type-209 diesel-electric submarines, designated the KSS-I *Changbogo*-class, which were subsequently built in South Korea under license. These are now being replaced by the German-designed Type-214 *Chungji*-class (KSS-II). The Type-214 is notable for being outfitted with fuel cells for air-independent propulsion, permitting the boat to remain submerged much longer (up to three weeks) than can conventional diesel-electric submarines. In 2000 South Korea ordered three Type-214 submarines to be built in Korea under license, the first of which was commissioned by the ROKN in 2008; in 2008 Seoul exercised an option to build six more boats, for a total of nine.[30] In addition, it is believed that South Korea will eventually design and build its own class of submarine, the KSS-III (perhaps up to nine boats).[31] Depending on how many KSS-IIs and KSS-IIIs are ordered, the ROKN could be operating a fleet of up to 18 modern diesel-electric submarines by 2020–25.

After protracted negotiations, India signed a contract in 2005 to acquire six Franco-Spanish *Scorpène*-type submarines (designated the *Kalvari*-class), which are currently constructed under license at India's Mazagon Docks shipyard; the final two in this batch are being outfitted with the French MESMA (*module d'energie sous-marine autonome*) system for air-independent propulsion; additional *Scorpène* production is anticipated.[32] In addition, the country is keen to develop a nuclear submarine fleet, and it is currently leasing two *Akula*-class submarines from Russia. India also wants to build its own nuclear-powered submarines – both hunter-attack (SSN) and ballistic missile-carrying (SSBN). Initially dubbed the Advanced Technology Vessel (ATV), this program was in development for more than 30 years. The Indian Navy (IN) launched its first nuclear-powered submarine, the *Arihant*, in 2009, and it was officially commissioned only in 2016. The IN plans to ultimately deploy a fleet of three SSBNs, armed with the indigenously developed *Sagarika* submarine-launched ballistic missile.[33]

Turning to Southeast Asia, Singapore possessed no submarine fleet at all until the late 1990s, when the RSN acquired four used 1960s-era submarines from Sweden. In 2009 Singapore took delivery of two more former Swedish Navy submarines; significantly, these boats, renamed the *Archer*-class, have been retrofitted with the Stirling AIP engine.[34] This was followed by a 2013 order for two brand-new Type-218SG submarines from the German shipbuilder Thyssen-Krupp Marine Systems; these boats will be delivered to the RSN by 2020. For its part, Kuala Lumpur has recently taken delivery of two Franco-Spanish *Scorpène*-class submarines for the Royal Malaysian Navy (RMN), both of which were commissioned in 2009, while the Vietnamese navy in 2010 ordered six *Kilo*-class diesel-electric submarines from Russia at a cost of US$2 billion.[35] In the mid-2000s rumors were floating around that Indonesia would acquire four *Kilo*-class and two *Lada*-class submarines from Russia to replace the navy's two aging (and probably inoperable) German-built Type-209 boats. However, this deal apparently fell through after Moscow refused to allow Jakarta to use Russian credits to construct a submarine base. In 2012, however, Jakarta signed a deal with South Korea for three new Type-209 submarines, with deliveries by 2018. Thailand also has a requirement for two or more submarines; ex-ROK Navy Type-209 boats have been mooted.

Airpower developments

Nearly every air force in the Asia-Pacific region currently possesses or is in the process of acquiring at least some fourth-generation or "fourth-generation-plus" fighter aircraft, capable of firing stand-off active radar-guided medium-range air-to-air missiles or delivering precision-guided air-to-surface munitions. China, for instance, has acquired approximately 300 Su-27 and Su-30 combat aircraft, including licensed-production of the Su-27 (dubbed the J-11B) at the Shenyang Aircraft Company in Manchuria. Moreover, the military is supplementing these purchases with the manufacture of its first indigenous fourth-generation-plus fighter, the J-10. The J-10 is an agile fighter jet in roughly the same class as the F-16C and features a delta wing/canard design, fly-by-wire flight controls, a "glass cockpit" (i.e., the use of liquid-crystal multifunction displays instead of analogue instruments), and, in later "B" versions, an active electronically scanned array (AESA) radar; over 400 J-10A/B fighters have been delivered to the PLA Air Force (PLAAF) since the mid-2000s.

By the late 2000s Japan had completed acquisition of approximately one hundred indigenous F-2 fighters (a heavily modified version of the F-16), to complement its force of over 200 F-15s. Over the next two decades Japan's Air Self Defense Force (ASDF) intends to acquire, either through import or indigenous development, a fifth-generation fighter to replace increasingly obsolescent F-4Js and older F-15s in ASDF inventories. In December 2011 Japan placed an order for 42 F-35 Joint Strike Fighters; these are currently being assembled in Japan and will be delivered to the ASDF starting in 2016. At the same time Japan has been quietly working on an indigenous fifth-generation fighter aircraft of its own, i.e., the X-2 ATD-X (Advanced Technology Demonstrator – Experimental). The X-2 is "a testbed platform for multiple technologies," including next-generation electronically scanned array radar, multi-dimensional thrust vectoring, an indigenous low-bypass turbofan engine, and radar-absorbing composite materials.[36] So far, the ATD-X has cost around 39.4 billion yen (around US$331 million) in development costs, and it had its first flight in early 2016. Production of an "F-3" fighter, based on the X-2, will not begin until 2027 at the earliest.

With regard to South Korea, the ROK Air Force (ROKAF) has acquired over 170 F-16C/D and 61 F-15K combat aircraft, and in 2015 it signed a contract for 40 F-35 JSFs. The ROKAF also operates at least 20 FA-50s, a light attack aircraft based on the indigenously developed T-50 trainer jet. More critically, the local aerospace industry was awarded a contract from the Korean government in 2015 to undertake development of a new indigenous combat aircraft, the KFX (Korean Fighter Experimental). The KFX will have stealthy features similar to the US-built F-35, as well as featuring an advanced radar and avionics. The ROKAF plans to procure as many as 120 KFX fighters, which will replace its aging fleet of F-4 Phantoms and F-5s.

India has obtained, either through import or licensed-production, 240 Su-30MKI fighter jets, and could eventually buy up to 50 more.[37] India also plans to acquire up to 220 indigenous *Tejas* Light Combat Aircraft, although it should be noted that the LCA program has been delayed for several years and only reached initial operating capability with the Indian Air Force in early 2015. In addition, in early 2016 New Delhi placed an off-the-shelf order for 36 French-built *Rafale* fighters, worth nearly US$9 billion.[38]

The Republic of Singapore Air Force (RSAF) is the most advanced of all Southeast Asian air forces. The RSAF, for example, possesses 74 F-16s of the latest Block 52/52+ type. In addition, in 2005 the RSAF placed its first order of F-15SG fighters, eventually acquiring a total of 32 aircraft, including a dozen F-15s that are stationed in the United States for RSAF training. Singapore is a partner in the international JSF program and could buy upwards of 100 F-35 fighters.[39] Other recent Southeast Asian fighter jet purchases include 18 Su-30MKM Flankers by Malaysia, with plans to buy another 18 fighter aircraft (either additional Su-30s or some other aircraft, such as the Swedish-produced *Gripen* or the US F/A-18); 12 *Gripens* by Thailand; and 16 Sukhois (5 Su-27s and 11 Su-30s) by Indonesia, with hopes to eventually purchase up to 40 additional Su-27/-30 aircraft.[40]

Missile defenses

In addition, several Asian-Pacific nations, including Japan, China, India, and Taiwan, are all in the process of acquiring missile defenses. Japan, for example, has recently completed upgrading its fleet of six *Aegis*-type destroyers to the US Navy's Sea-based Midcourse Defense (SMD) missile defense mode. The SMD upgrade entails improvements to the current SPY-1 multifunction phased-array radar and fire-control system that increase the range and altitude of its search, detection, track, engagement, and control functions in order to handle exo-atmospheric anti-missile engagements. This program also entails the deployment of a new interceptor missile, the Standard SM-3 Block IA missile, which includes a third-stage for extended range and a Lightweight Exo-Atmospheric Projectile (LEAP) kinetic warhead for terminal homing and intercept. Japan's SMD system should be fully deployed by 2011. SMD is complemented by the land-based Patriot PAC-3 system that provides endo-atmospheric protection against missile threats to the Japanese homeland.

Other Asia-Pacific nations are following suit with their own missile defense plans. India has purchased the Israeli Green Pine ballistic missile early-warning radar, and New Delhi is currently working to create a national missile defense system that uses both the Russian S-300 surface-to-air missile and a variety of indigenously developed exo-atmospheric and point-defense missile systems. China conducted a missile defense test in early 2010, and Taiwan is attempting to modify its indigenous *Tien Kung* II SAM into a working missile interceptor.[41] Korea's planned acquisition of several *Aegis*-equipped warships could conceivably provide the basis for their national missile defenses based on the SMD concept. Seoul recently announced plans to inaugurate an indigenous missile defense system during the 2010s in order to defend against North Korean ballistic missile threats. This program will likely include both land- and sea-based inceptors and will cost at least 300 billion won (US$214 million).[42] Finally, South Korea has agreed to the deployment of the Terminal High Altitude Area Defense (THAAD) by US forces deployed on the peninsula.

Long-range, precision-strike weapons

At least as important as the acquisition of modern military platforms throughout the Asia-Pacific is the steady proliferation of precision-guided weapons for stand-off strike. As mentioned already, many new surface combatants and submarines being deployed in the region are equipped with advanced anti-ship cruise missiles, such as the Harpoon on Singapore's Formidable-class frigate and on Japan's *Sōryū*-class submarine; the Exocet on Malaysian and Indian *Scorpène*-class submarines; and the Russian 3M-80E *Moskit* on Chinese *Sovremennyy*-class destroyers. India has developed the *Brahmos* supersonic anti-ship cruise missile in cooperation with Russia, which will be deployed in a variety of sea-, land-, and air-based modes, and Taiwan is currently developing the *Hsiung Feng* III (HF-3) supersonic ASCM.

Additionally, many countries are acquiring active radar-guided, medium-range air-to-air missiles (AAMs) for their fighter aircraft. These include the US AMRAAM (advanced medium-range air-to-air missile) by Japan, South Korea, Singapore, and Thailand; the Russian R-77/AA-12 by China, Indonesia, and Malaysia; and the PL-12 by China. In the case of AMRAAM, this missile was embargoed for sale to many states in the region until recently.

At the same time, Asian-Pacific militaries are being increasingly equipped with stand-off land-attack munitions. Japan, South Korea, and Singapore are buying the GPS-guided Joint Direct Attack Munition (JDAM), while Singapore also is also acquiring the Joint Stand-Off Weapon (JSOW), a precision-guided glide bomb with a range of up to 130 kilometers. More importantly, perhaps, several countries in the region have acquired several types of land-attack cruise missiles, many of them adapted from existing ASCMs. Taiwan, for example, is deploying the *Hsiung Feng* IIE (HF-2E) LACM, based on its HF-2 anti-ship missile, China has developed the *Dong-Hai* 10 (DH-10) LACM, and South Korea has developed the *Hyunmoo*-IIIC LACM.

Finally, it is important not to discount the strike value of ballistic missiles armed with non-nuclear warheads. China, of course, has deployed a large number of conventionally armed surface-to-surface missiles, including the 300 kilometer-range

DF-11 (CSS-7) and the 600 kilometer-range DF-15 (CSS-6) short-range missiles. This is in addition to China's growing arsenal of sophisticated long-range nuclear-tipped ballistic missiles, including the DF-31 (CSS-9) road-mobile, solid-fuel intercontinental ballistic missile (ICBM), with a range of 8,000 kilometers, and the submarine-launched JL-2 (CSS-N-4) missile. India, meanwhile, has developed the short-range *Prithvi* and medium-range *Agni* missiles as nuclear delivery vehicles, and it is currently field-testing an SLBM. North Korea, of course, has deployed its notorious *Nodong*-1 medium-range ballistic missile, and the intermediate-range (6,000 kilometer) *Taepodong*-2 is under development. Other tactical missile systems in use in the region include the MGM-140 Army Tactical Missile System (ATACMS) in service with the South Korean army, the U.S.-built HIMARS multiple rocket launcher system in Singapore, and the Brazilian ASTROS-II artillery rocket in Malaysia.

Command, control, communication, computing, intelligence, reconnaissance, and surveillance (C4ISR)

Finally, many Asia-Pacific militaries are engaged in greatly expanding and upgrading their capabilities for C4ISR.[43] For example, China, Japan, Singapore, and Taiwan all currently possess airborne early warning and command (AEW+C) aircraft, while India, South Korea, and Thailand intend to acquire AEW+C aircraft in the near future. Both Japan and South Korea have the *Aegis* naval sensor and combat system deployed on their largest surface combatants, and Taiwan has acquired long-range early-warning radar for ballistic missile detection and tracking from the United States.

In addition, nearly every major military in the region is acquiring unmanned aerial vehicles (UAV), while China, India, Japan, South Korea, and Taiwan have all launched satellites for surveillance, communications, or navigation/target acquisition. Moreover, these countries and others in the region are also able to exploit imagery provided by a host of commercial earth-observation satellite operators, such as IKONOS, EROS, and QuickBird.

Several countries in the region – in particular, China, Japan, Singapore, South Korea, and Taiwan – have also made or are presently making considerable investments in new types of information processing, command and control, and communications and datalinks. The South Korean military, for its part, is developing an integrated tactical communications system, while Taiwan is spending billions of dollars on a new military-wide C4ISR network that will link sensors, computers, and communications across the services.[44] The Singapore Armed Forces already possesses a secure C4I network, utilizing microwave and fiber-optic channels and linked to air and maritime surveillance systems. In addition, as part of its "Integrated Knowledge-based Command and Control" (IKC2) concept Singapore is putting considerable focus on expanding its capabilities for network-centric warfare.[45]

China in particular has put considerable emphasis on upgrading C4ISR assets according to its concepts of "informationalized" warfare. Consequently, the PLA is expanding the use of satellites for communication, surveillance, and navigation, exploiting its manned space program for military purposes, and greatly expanding its use of cyberspace for offensive operations. Moreover, the PLA has invested considerable resources in creating a separate military communications network that uses fiber-optic cable, satellites, microwave relays, and long-range high-frequency radio. The PLA has also particularly focused on developing its capacities for cyber war, including computer network attacks (that is, disrupting the enemy's computer networks), electronic warfare, and even physical attacks on the enemy's C4SIR network (such as with antisatellite [ASAT] weapons).[46]

Assessing the impact of modernization on military capabilities

The arms build-up in the Asia-Pacific over the past 15 years or so is undeniably significant. In the first place, recent acquisitions by regional militaries constitute something more than mere modernization; rather, the new types of armaments being procured and deployed promise to significantly affect regional war-fighting capabilities. Local militaries are acquiring greater lethality and accuracy at longer ranges. Stand-off precision-guided weapons – such as land-attack cruise missiles, tactical ballistic missiles, and a variety of smart munitions, some carried by fourth-generation-plus fighter aircraft – have greatly increased these militaries' firepower and effectiveness, making them capable of longer distance and yet more precise attack. Additionally, militaries in the Asia-Pacific are acquiring new or increased capabilities for force projection, operational maneuver, and speed. Modern submarines and surface combatants, amphibious assault ships, aircraft carriers, air-to-air refueling abilities, and transport aircraft have all extended these militaries' theoretical range of action. Regional militaries are also more survivable due to the increased use of stealth and active defenses, particularly missile defense. Finally, these forces are improving their capabilities for battlefield knowledge, situational awareness, and command and control. New platforms for reconnaissance and surveillance, especially in the air and in space, have considerably expanded these militaries' capacities to look out over the horizon and across all five areas of the future battlespace: ground, sea, air, space, and cyber.

More importantly, many Asia-Pacific militaries are acquiring the types of military equipment that could fundamentally transform their forces along the lines of the information-technologies (IT)-led "revolution in military affairs" (RMA). This embrace of network-enabled warfare – known in China as "winning wars under conditions of informatization," and in Singapore as the IKC2 concept – is a potentially historic shift. Regional militaries could be on the cusp of bundling together sensors, computers, communications, command and control systems, munitions, and platforms that would greatly improve the synergy of their fighting effectiveness. Such emerging capabilities, particularly on the part of China, could in turn greatly affect strategy and operations in future military endeavors in the Asia-Pacific.

Certainly most Asia-Pacific militaries in the 21st century are a vast improvement over their predecessors of 20 or even 15 years ago, given the addition of fourth-generation-plus combat aircraft, new classes of warships and submarines, precision-strike weapons, and so on. In China, for example, the J-10 and Su-30 fighters have replaced MiG-19s and MiG-21s. Likewise, F-15s are replacing F-4s in the ROKAF and A-4s in the RSAF, and India is supplementing vintage Jaguars, MiG-27s, and Mirage-2000s with Su-30s and the *Tejas* LCA. Additionally, beyond-visual-range, active radar-guided air-to-air missiles, such as the AMRAAM and AA-12, are replacing or supplementing older generation AAMS, such as the short-range AIM-9 Sidewinder or the semi-active AIM-7 Sparrow. Moreover, Japan and South Korea have both signed contracts to acquire the F-35 fifth-generation fighter, and Singapore and perhaps India are also potential customers for the JSF. In terms of surface combatants, countries such as China, India, Japan, and South Korea are acquiring advanced destroyers with sophisticated radars, surface-to-air missiles, and combat systems that provide their militaries with long-range air defense at sea – and even missile defense – capabilities that they did not earlier possess. In the past 15 years, countries such as South Korea, Malaysia, Singapore, and Vietnam, which never possessed much in the way of submarines forces, or, indeed, any submarines at all, are being equipped with modern boats. In the case of Japan, India, the ROK, and Singapore, these submarines are outfitted with air-independent propulsion that permits them to remain submerged for much longer periods of time. China and India, for their part, have highly ambitious nuclear-powered submarine (both SSN and SSBN) programs. Finally, many Asia-Pacific militaries are being equipped for the first time with a variety of stand-off precision-strike weapons, including JDAM (Japan, South Korea, and Singapore), JSOW (Singapore), and the AGM-142 air-to-surface missile (South Korea). Just as importantly, South Korea and Taiwan have developed their own land-attack cruise missiles, while China and India have gained new capabilities for using ballistic missiles as battlefield strike weapons. In addition, these forces are certainly better equipped than in the past with systems for communications, command and control, intelligence, and surveillance. For example, China, India, Japan, and Singapore (and soon South Korea) have all acquired airborne early-warning and command aircraft, while UAVs have proliferated throughout the region.

That the Asia-Pacific nations have added considerably to their military arsenals is not in doubt. Nor does the process of military modernization – propelled by regional geopolitical forces, enabled by robust defense spending and a buyer's market in the international arms market, and stirred by the transformative promise of network-centric warfare – seem to show any signs of abating. Moreover, countries in the region are acquiring hardware that, on the surface at least, imbues their militaries with new capacities for war-fighting when it comes to mobility, speed, precision-strike, firepower, battlespace intelligence, and cyber.

Just as important, however, many nations in the Asia-Pacific seek to supplant, or at least supplement, foreign arms suppliers with indigenous producers of needed weapons systems. The ongoing process of regional military modernization has been, and increasingly will be, marked by a growing preference for homegrown weaponry for the reasons laid out in Chapter 1. The emphasis, therefore, is increasingly on the acquisition of advanced armaments *and* on the indigenous sourcing of said armaments. It is to addressing the individual histories, experiences, and future prospects of regional arms manufacturers that this book next turns.

Notes

1 "Section IV: Building and Development of China's Armed Forces," *China's Military Strategy* (*Beijing*: Ministry of National Defense, May 26, 2015). See also Dennis J. Blasko, "The 2015 Chinese Defense White Paper on Strategy in Perspective: Maritime Missions Require a Change in the PLA Mindset," *China Brief*, May 29, 2015.
2 "Section IV: Building and Development of China's Armed Forces," *China's Military Strategy*. (*Beijing*: Ministry of National Defense, May 26, 2015).
3 "Section II: National Defense Policy," *China's National Defense in 2006* (Beijing: Information Office of the State Council of the People's Republic of China, December 29, 2006).
4 Anthony H. Cordesman and Steven Colley, *Chinese Strategy and Military Modernization in 2015: A Comparative Analysis* (Final Review Draft) (Washington DC: Center for International and Strategic Studies, October 10, 2015), p. 41.
5 "Section IV: Building and Development of China's Armed Forces," *China's Military Strategy* (*Beijing*: Ministry of National Defense, May 26, 2015).
6 "Section II: National Defense Policy," *China's National Defense in 2004* (Beijing: Information Office of the State Council of the People's Republic of China, 2004).
7 OSD, *Annual Report 2007*, pp. 5, 11–14.
8 "Section III: Revolution in Military Affairs with Chinese Characteristics," *China's National Defense in 2004*; Office of the Secretary of Defense (OSD), *Annual Report to Congress: Military and Security Developments Involving the People's Republic of China 2011* (Washington DC: Office of the Secretary of Defense, 2011), p. 22.
9 "Section IV: Building and Development of China's Armed Forces," *China's Military Strategy* (*Beijing*: Ministry of National Defense, May 26, 2015).
10 "Section IV: Building and Development of China's Armed Forces," *China's Military Strategy* (*Beijing*: Ministry of National Defense, May 26, 2015). See also Joe McReynolds, "Network Warfare in China's 2015 Defense White Paper," *China Brief*, June 19, 2015.
11 Rahul Bedi, "Getting in Step: India Country Briefing," *Jane's Defense Weekly*, February 6, 2008.
12 Bedi, "Getting in Step."
13 Geoffrey Till, *Globalization: Implications of and for the Modern/Post-modern Navies of the Asia Pacific*, RSIS Working Paper No. 140 (Singapore: S. Rajaratnam School of International Studies, October 15, 2007), p. 37.
14 Till, *Globalization*, p. 37.
15 Till, *Globalization*, p. 38.
16 Till, *Globalization*, p. 37.
17 Greg Torode, "Exercises Off Japan and Taiwan Show PLA Navy's New Strength," *South China Morning Post*, April 18, 2010.
18 Hoon Noh, *South Korea's 'Cooperative Self-Reliant Defense': Goals and Directions*, KIDA Paper No. 10 (Seoul: Korea Institute for Defense Analyses, April 2005), p. 5; and the Republic of Korea, Ministry of National Defense, *2004 Defense White Paper* (Seoul: Ministry of National Defense, 2004), 92–98.
19 See Ian Storey, "China's 'Charm Offensive' Loses Momentum in Southeast Asia [Part I]," Jamestown Foundation, *China Brief*, April 29, 2010; and Ian Storey, "China's 'Charm Offensive' Loses Momentum in Southeast Asia [Part II]," Jamestown Foundation, *China Brief*, May 13, 2010.

20 Andrew Tan, *Force Modernization Trends in Southeast Asia*, Working Paper, 2004, Institute of Defence and Strategic Studies, pp. 30–31.
21 Stockholm International Peace Research Institute (SIPRI), SIPRI Military Expenditure Database, 2010 (http://milexdata.sipri.org).
22 SIPRI Military Expenditure Database.
23 SIPRI Military Expenditure Database.
24 "LP-X Dokdo (Landing Platform Experimental) Amphibious Ship," GlobalSecurity. org (http://www.globalsecurity.org/military/world/rok/lp-x.htm).
25 "LP-X Dokdo."
26 Bedi, "Getting in Step."
27 "The Vikramaditya [ex-Gorshkov] Aircraft Carrier," GlobalSecurity.org (http://www.globalsecurity.org/military/world/india/r-vikramaditya.htm).
28 Edward Hooten, "Modernizing Asia's Navies," 8; and "The Vikrant-class Air Defense Ship," GlobalSecurity.org (http://www.globalsecurity.org/military/world/india/ads.htm).
29 Bedi, "Getting in Step."
30 Tim Fish, "Seoul Commissions Type 214 Sub," *Jane's Defense Weekly*, January 23, 2008.
31 Robert Karnoil, "Team Prepares for 2007 Start on KSS-III Design," *Jane's Defense Weekly*, December 20, 2006.
32 Rajat Pandit, "India Plans to Buy 6 New Subs, Says Navy Chief," *Times of India*, December 2, 2007.
33 Sandeep Unnithan, "The Secret Undersea Weapon," *India Today*, January 17, 2008.
34 Tim Fish and Richard Scott, "Archer Launch Marks Next Step for Singapore's Submarine Force," *Jane's Defense Weekly*, June 18, 2009.
35 Nga Pham, "Vietnam to Buy Russian Submarines," *BBC News*, December 16, 2009.
36 Franz-Stefan Gady, "Japan Unveils News Fifth-Generation Stealth Fighter Jet," *Diplomat*, January 29, 2016
37 Rajat Pandit, "IAF Wants 50 More Sukhois to Counter China, Pakistan," *Times of India*, October 2, 2009.
38 Vivek Raghuvanshi, "India and France to Finalize $8.9 Billion Deal for 36 Rafales," *Defense News*, April 19, 2016.
39 Jermyn Chow, "F-15 Training Cements Ties with US," *Straits Times*, November 21, 2009.
40 Trefor Moss, "Painful Progress: Indonesia Country Briefing," *Jane's Defense Weekly*, October 16, 2009; and Tan, "Force Modernization Trends in Southeast Asia," 17.
41 "China: Missile Defense System Test Successful," *USA Today*, January 11, 2010; and "Taiwan to Upgrade to Tien Kung-2 SAM," Missilethreat.com, July 31, 2006 (http://www.missilethreat.com/archives/id.419/detail.asp).
42 "South Korea to Complete Missile Defense by 2012," *Defense News*, February 15, 2010.
43 Jason Sherman, "Digital Drive: Focus, Funding Shifts to C4ISR, Precision Weaponry," *Defense News*, February 16, 2004, pp. 23–24.
44 Sherman, "Digital Drive"; and Jason Sherman, "Taiwan to Build Military-Wide C4ISR Network," *Defense News*, October 7, 2003.
45 Bernard Fook Weng Loo, "Transforming the Singapore Armed Forces: Problems and Prospects" (paper presented at the conference Defense Transformation in the Asia-Pacific: Meeting the Challenge, Honolulu, March 30–April 1, 2004), p. 5; and Tim Huxley, "Singapore and Military Transformation" (paper presented at conference The RMA for Small States: Theory and Application, Singapore, February 25–26, 2004), p. 2.

46 US Department of Defense, *Annual Report on the Military Power of the People's Republic of China 2009* (Washington, DC: US Department of Defense, 2009), pp. 25–28; You Ji, "China's Emerging National Defense Strategy," Association for Asian Research, January 12, 2005; Wendell Minnick, "China Shifts Spending Focus to Info War," *Defense News*, September 11, 2006; and Bill Gertz, "Inside the Ring: China Info Warfare," *Washington Times*, June 2, 2010; Richard Bitzinger, "China's RMA: Good Enough?" *International Relations and Security Network*, August 29, 2007 (http://www.isn.ethz.ch/isn/Current-Affairs/Security-Watch/Detail/?ots591=0C54E3B3–1E9C-BE1E-2C24-A6A8C7060233&lng=en&id=53705).

3 China

The People's Republic of China (PRC) has, going back to its founding, strived to become self-reliant in the development and manufacture of arms.[1] The current phrase to express this desire for autarky in defense production and acquisition is, according to Tai Ming Cheung, *you zhongguo tese de zizhu chuangxin*, or "innovation with Chinese characteristics."[2] Bates Gill and Taeho Kim argue that China's desire to wean itself off foreign dependencies for armaments actually goes back more than 150 years, when the country was too weak militarily to fend off encroaching Western powers.[3] For most of its history, however, the results of these endeavors have been decidedly mixed. Even with sizable economic inputs, access to foreign technologies, and considerable political will, China, up until the late 1990s, experienced only limited success when it came to the local design, development, and manufacture of advanced conventional weapons. Most systems were at least a generation or two behind comparable military equipment being produced at the time in the West or in the Soviet Union, and problems with quality and reliability abounded. In addition, overcapacity, redundancy, inefficient production, and above all a weak defense research and development (R&D) base all conspired to impede the development of an advanced indigenous arms production capability. Overall, these circumstances left China in the unenviable position of pursuing great power status with a decidedly "Third World" arms industry.

Not surprisingly, therefore, *reforming* the local defense industry in order to upgrade its technology base and manufacturing capabilities and to make armaments production more efficient and cost-effective has long preoccupied the Chinese leadership. The fact that most of these efforts had little positive impact on the country's military technological and industrial capabilities only encouraged Beijing to experiment with additional reforms in the hopes of finally getting it right. The most recent round of defense industrial base restructuring and reorganization began in the late 1990s and basically entailed a concerted effort to inject more market-oriented notions – along with additional funding and new technologies, both foreign and commercial – into the Chinese military-industrial complex.

These modernization efforts have had several objectives. For one thing, as China strives to become a global power, it is increasingly seeking "hard" power, i.e., military strength, commensurate with its growing economic, diplomatic, and cultural "soft" power. Second, Beijing appears to be more prone to use military force (or the threat of military force) to defend and promote its regional interests, such as its territorial claims in the South China Sea; consequently, building up that military wherewithal is instrumental to this strategy. Third, China aspires to increase its military capacities in order to keep the pressure on Taiwan *not* to declare independence *and* to eventually accept some kind of reunification with the mainland; at the same time, China wants to reduce or eliminate the capabilities of the United States to intervene on behalf of Taiwan in case of a cross-straits military clash. Finally, China overall seeks military power to mitigate the rising American military presence in the Asia-Pacific and to establish itself as a credible rival to the United States in this region.

For whatever reason, these efforts have paid remarkable dividends, and since the late 1990s the People's Liberation Army (PLA) has made amazing progress in transforming itself into a modern fighting force; in many instances it is practically unrecognizable when compared to the PLA that existed before 1997. The impact of this transformation has been particularly noticeable in the past few years in the form of a much more assertive, even aggressive, China that is increasingly willing to use its military to protect and advance its national interests. Prominent examples of this increased use of the PLA as an instrument of national policy include the dispatch of PLA Navy (PLAN) vessels to fight piracy in the Gulf of Aden and the PLAN's recent launching of an aircraft carrier. What the end result of this military modernization process will be, or how China may further use its growing military power, is still an open question.

1997 is a good place to start when addressing the current modernization of the PLA, as this was a watershed year in the history of the Chinese military. It was around this time, for instance, that the PLA officially adopted the strategic concept of fighting "limited local wars under high-technology conditions" (and later, "under conditions of informatization"), which still drives current operational and hardware requirements for military modernization. Starting that year, too, Chinese defense spending began its remarkable and near-unbroken run of double-digit real annual growth, which has underwritten the process of military modernization that was to follow. Also in 1997, the decision was made by the central government to force the PLA to divest itself of the bulk of its commercial activities so as to concentrate on its primary function, that is, deterrence, compellence, and, if necessary, war-fighting. Finally, and most important to this chapter, the 15th Party Congress in September 1997, the Chinese Communist Party (CCP), decided to radically reform the state-owned enterprise (SOE) sector, which marked the beginning of the current process of restructuring and upgrading the Chinese defense industry.

China's military-industrial complex up until the late 1990s

China possesses one of the oldest, largest, and most diversified military-industrial complexes in the developing world. The modern Chinese arms industry has its roots in the 1950–53 Korean War and the initiation of some 41 "key projects" for weapons production during the 1950s.[4] By the late 1990s the Chinese defense industrial base was an agglomeration of approximately 1,000 enterprises (each comprising multiple factories, research units, trading companies, and schools and universities) and 200-plus major research institutes, employing some 3 million workers as well as 300,000-plus engineers and technicians.[5] In particular, China is one of the few countries in the developing world to produce a full range of military equipment, from small arms to armored vehicles to fighter aircraft to warships and submarines, in addition to nuclear weapons and intercontinental ballistic missiles.

Up until the early 1990s armaments production was largely concentrated in a handful of machine-building ministries with responsibility for a particular defense sector (e.g., the Ministry of Aviation and Aerospace Industry, which manufactured aircraft and missile systems). In 1993, in an effort to "corporatize" the defense industrial base, these ministries were reorganized into five state-owned enterprises (SOEs): Aviation Industries of China (AVIC), China Aerospace Corporation (CASC), China Ordnance Industry Corporation (also known as NORINCO), China State Shipbuilding Corporation (CSSC), and China National Nuclear Corporation (CNNC). Subsequently, their industrial responsibilities were:

- AVIC: fighter aircraft, bombers, transports, trainer jets, helicopters, attack aircraft, unmanned aerial vehicles (UAVs).
- CASC: tactical and strategic missile systems, space-launch vehicles, satellites.
- NORINCO: all ground ordnance, including tanks, armored vehicles, artillery systems, small arms and ammunition
- CSSC: all warships, including destroyers, frigates, submarines, corvettes, patrol boats.
- CNNC: nuclear weapons, nuclear energy development, nuclear power plants, nuclear fuel and equipment

The Commission on Science, Technology, and Industry for National Defense (COSTIND) nominally administered these five SOEs, while COSTIND also directed the country's military R&D program. Key Chinese weapons systems produced before the late 1990s include the J-6, J-7, and J-8II fighters (the J-6 and J-7 were, respectively, licensed-produced versions of the Soviet MiG-19 and MiG-21 fighters), the JH-7 fighter-bomber, the Silkworm and C-802 anti-ship cruise missiles, the 600 km-range DF-15 (M-9) and the 300 km-range DF-11 (M-11) surface-to-surface missiles (SSMs), and the Type-035 *Ming*-class submarine (a licensed-produced version of the Soviet *Romeo*-class).

Despite these accomplishments, the Chinese military-industrial complex suffered from a number of shortcomings that in turn inhibited translating breakthrough technologies and designs into comparatively advanced weapon systems. As late as the late 1990s China still possessed one of the most technologically backwards defense industries in the world; most indigenously developed weapons systems were at least 15 to 20 years behind those of the West – basically comparable to 1970s- or (at best) early 1980s-era technology – and quality control was consistently poor. China's defense research and development base was regarded to be deficient in several critical areas, including aeronautics, propulsion (such as jet engines), microelectronics, computers, avionics, sensors and seekers, electronic warfare, and advanced materials. Furthermore, the Chinese military-industrial complex has traditionally been weak in the area of systems integration.[6]

Consequently, aside from a few "pockets of excellence," such as ballistic and anti-ship missiles, the Chinese military-industrial complex by the late 1990s appeared to demonstrate few capacities for designing and producing relatively advanced conventional weaponry. Especially when it came to combat aircraft, surface combatants, and ground equipment, the Chinese generally confronted considerable difficulties when it came to moving prototypes into production, resulting in long development phases, heavy program delays and low production runs. The J-10 fighter jet – China's premier fourth-generation-plus combat aircraft – took more than a dozen years to move from program start to first flight, for example, and nearly 20 years before it entered operational service with the PLA Air Force (PLAAF).[7] Even after the Chinese begin building a weapon system, production runs were often small and fitful. According to Western estimates, during much of the 1990s the entire Chinese aircraft industry of around 600,000 workers manufactured only a dozen or so fighter aircraft a year, mainly 1960s- and 1970s-vintage J-8IIs and J-7.[8] According to the authoritative *Jane's Fighting Ships*, China launched only three destroyers and nine frigates between 1990 and 1999 – a little more than one major surface combatant per year. Moreover, the lead boat in the *Song*-class submarine program – China's first indigenously designed diesel-electric submarine – was only commissioned in 1999, eight years after construction began.[9]

Consequently, despite years of arduous efforts the inability of China's domestic defense industry to generate the necessary technological breakthroughs for advanced arms production meant that Beijing continued to rely heavily – even increasingly – upon direct foreign technology inputs in critical areas. The J-10 fighter, for example, was heavily based on technology derived from Israel's cancelled *Lavi* fighter jet program. These foreign dependencies are especially acute when it comes to jet engines, marine diesel engines, and fire-control radar and other avionics. For example, endemic "technical difficulties" surrounding the JH-7 fighter-bomber's indigenous engine resulted in significant program delays, forcing the Chinese to approach the British in the late 1990s about acquiring additional Spey engines in order to keep the aircraft's production line going; additionally, current versions of the J-10 are being outfitted with a Russian engine until the Chinese aviation industry is able to perfect an indigenous replacement.[10] Chinese surface combatants incorporate a number of foreign-supplied systems, including Ukrainian gas turbine engines, French surface-to-air missiles, Italian torpedoes, and Russian ship-based helicopters.

In fact, until the late 1990s and even the early 2000s the sharpest edges of the pointy end of the PLA spear were still mostly of foreign – and particularly

Russian – origin. From this, one may infer that the Chinese military was still dissatisfied with the quality and capabilities of weapon systems coming out of domestic arms factories, or that local arms manufacturers remained unable to produce sufficient numbers of the kinds of weapons that the PLA wants in the near future. Consequently, from the middle of the 1990s to the middle of the first decade of the 21st century China was one of the world's largest arms importers, and between 1998 and 2005 Beijing signed new arms import agreements worth some US$16.7 billion; as recently as 2005 it purchased US$2.8 billion worth of foreign weapon systems.[11] In the early 1990s, for example, despite the fact that China already had four fighter aircraft programs either in production or development – the J-7, J-8II, JH-7, and J-10 – the PLA nevertheless decided to buy several dozen Su-27 fighters; this purchase was later supplemented by an agreement to license-produce 200 Su-27s and then a subsequent purchase of approximately 100 more advanced Su-30 fighter-bombers. The PLA Navy also acquired 12 *Kilo*-class submarines and 4 *Sovremennyy*-class destroyers (armed with supersonic SS-N-22 anti-ship cruise missiles), even though Chinese shipyards are building the *Song*-class submarine and several new types of destroyers. Other Russian-supplied weapons systems included S-300 and SA-15 surface-to-air missiles, Il-76 transport planes, airborne early warning aircraft, AA-12 active-radar guided air-to-air missiles, and various precision-guided munitions. With few exceptions (such as tactical ballistic missiles or nuclear submarines), foreign weapons systems remained the most critical force multipliers when it came to calculating Chinese military power. Therefore, autarky, or self-sufficiency in armaments production, continued to elude the PLA – in some cases, quite significantly.

Compounding these technological deficiencies and continuing foreign dependencies were a number of structural, technical, and "corporate cultural" deficiencies that impeded the design, development, and manufacture of advanced conventional weapons. Overall, arms production in China had largely been, for several decades, an inefficient, wasteful, and unprofitable affair. One reason was excess capacity: quite simply, China possessed far too many workers, too many factories, and too much productive capacity for what few weapons it produced, resulting in redundancy and a significant duplication of effort, inefficient production, and wasted resources. The Chinese aircraft industry, for example, was estimated in the late 1990s to possess a workforce nearly three times as large as it required.[12] Within the shipbuilding industry, output during the same time period was only 17 tons per person per year, compared to around 700 tons per person in shipyards in more advanced countries.[13]

By the mid-1990s as well, at least 70 percent of China's state-run factories were thought to be operating at a loss, and the arms industries were reportedly among the biggest money losers. As a result, most defense firms were burdened with considerable debt, much of which was owed to state-run banks (who were obliged to lend money to state-owned firms); at the same time, arms factories were owed money by other unprofitable state-owned companies, which was nearly uncollectible.[14]

The creation of China's "Third Line" defense industries – that is, the establishment of redundant centers of armaments production in the remote interior of southern and western China – in the 1960s and 1970s only exacerbated this overcapacity, underutilization, and unprofitability of the Chinese military-industrial complex.[15] Estimates are that from 1966 to 1975 Third Line construction consumed perhaps two-thirds of all industrial investment. Even by the late 1990s, approximately 55 percent of China's defense industries were located within the Third Line, yet most of these industries were much less productive than were coastal area factories and which continued operate in the red.[16]

Another structural impediment was the highly compartmentalized, vertically integrated, and secretive nature of the Chinese defense industry. Such a stovepiped and stratified environment made it difficult for arms producers to diffuse advanced technologies, gain access to advanced, militarily usable civilian technologies, share learning experiences, and collaborate on advanced weapons projects. In particular, it limited communications between R&D institutes that designed the weapons and the arms factories that produced these systems; between defense enterprises when it came to collaborating on weapons projects; and even between the defense industry and its major consumer, i.e., the PLA, when it came to requirements and specification. It exacerbated redundancy and the duplication of effort within the arms industry, as each defense enterprise tried to "do it all," resulting in the maintenance of expensive but underutilized manufacturing processes, such as dedicated second- and third-tier supplier networks and the establishment of in-house machine shops for parts production instead of outsourcing such manufacturing to other firms.

In addition to these structurally based barriers, China's defense industrial base for a long time lacked many of the basic technical skills necessary to fully exploit acquired technologies. One of the most critical of these was the defense industry's apparently enduring weak systems integration capabilities – the ability to envision, design, and develop a finished weapon system out of hundreds or even thousands of disparate components and subsystems and to get it to function to its fullest potential as a single unit. According to a US aviation industry representative, the Chinese were "especially deficient in systems integration" when it came to aircraft manufacturing; he added that "Chinese engineers and technicians are normally grounded in the basic discipline [of aircraft production], however, practical applications, manufacturing technologies, and overall experience are in short supply."[17] Pollack and Mulvenon, in their seminal (if unpublished) study of the Chinese aircraft industry in the mid-1990s, asserted that when it came to the systems integration and engineering process, "the Chinese do not have a master plan that builds aircraft from the bottom up. Instead, they try to take parts off the shelf that were never designed to be part of any particular end product and try to make them fit."[18] Even the controversial 1999 report by the US House of Representatives on Chinese military-technological capabilities – the so-called "Cox Report," named after Christopher Cox, who chaired the committee that undertook the study, and which accused China of stealing US military-technical secrets – conceded that China "lacks the ability to integrate the contributions of many disciplines that are required to utilize the rapidly emerging new technologies. The PRC system is unable to keep up with these basically new approaches."[19]

Even imports of foreign technologies did not always help the Chinese defense industry overcome this systems integration impediment. According to Mark Stokes, Chinese defense enterprises traditionally purchased foreign equipment "without much thought as to how to integrate various components" into a single, workable system.[20] As such, arms producers may have acquired the "know-how" but not necessarily the "know-why" of advanced weapons development and manufacturing.[21]

Closely linked to the problem of poor systems integration and systems engineering skills was limited workforce expertise on the part of industry. According to one Western analyst, the Chinese traditionally paid insufficient attention to training and workforce development.[22] Hence, advanced machine tools often went underused (or even unused), due to a lack of skilled operators.[23] In addition, factories often had so little actual contract work that many skilled workers gained only very limited experience with advanced manufacturing techniques; at the same time, many young, bright, and enthusiastic engineers and technicians had limited opportunities to apply their knowledge to actual programs.[24] For their part, SOE executives typically lacked the managerial and entrepreneurial skills and experiences necessary to make market-oriented investment and production decisions.[25]

China's military-industrial complex long functioned under a corporate culture that, in a manner typical of most state-owned enterprises, was highly centralized, hierarchical, bureaucratic, and risk-averse. This in turn stymied innovation, retarded R&D, and further added to program delays. In a 1999 study on Chinese capacities for innovation, two Western analysts argued that "Chinese managers do not have either the will, the expertise, or the freedom to take the risks and make the adjustment associated with innovations." Consequently, production management was often highly centralized and "personality-centric," with most critical project decisions being made by a single chief engineer. At the same time, lower-level managers tended to be "conformist, adhering to standard rules and procedures rather than to personal judgments based on their professional experiences." Hence, they were usually reluctant to make "learning mistakes" or to act on their own to deal with problems that might arise on the factory floor, thereby inhibiting experimentation and innovation.[26] Pollack and Mulvenon, for example, noted that Chinese technicians in the aviation industry often lacked practical experience in handling modern technology because of their "hesitancy of making a mistake and failing."[27]

This statist, risk-averse "SOE mindset" permeated throughout the defense sector, undermining competitiveness and the rationalization of arms production. For example, senior executives in China's military-industrial complex were often awarded contracts in order to preserve jobs and keep factories open, rather than on the basis of merit and competency.[28] For its part, the central government was traditionally quite forgiving of bad performance within the SOE sector.

Overall, regarding all of China's problems with armaments production up until the late 1990s, one US aerospace industry representative perhaps summed it up best:

> Part of the problem with Chinese [aircraft] manufacturing . . . is that industrial management in China still relies on 1950s Soviet styles. This involves "batch-building" a full order of aircraft in advance based on state-planned and dictated order for parts and materials. As a consequence of this system, there are no direct lines of accountability for quality control, and no cost-cutting discussions or steps available to mid-level management. There is no competitive bidding for contracts, workers are redundant, and schedules continually slip because state planning doesn't have a fixed required-delivery date for products . . . Young managers stay risk-averse and are reluctant to change or improve the system.[29]

Reforming China's defense industry, 1997 to the present

To be sure, the Chinese have long been aware of the deficiencies in their defense industrial base, and they have undertaken several rounds of reforms since the late 1980s in order to improve and modernize their military R&D and arms production processes. Most of these efforts fell well short of their intentions, however, because they failed to tackle the basic if endemic problems facing the defense industry: lack of competition, lack of accountability, excess capacity, lack of capital, lack of human skills, and a "statist" corporate culture. These prior failures make the post-1997 reform efforts even more significant, because they, more than earlier attempts, have tried to attack the very nature of Chinese arms development, production, and acquisition in order to, first, inject rational, requirements-based planning into the arms procurement process, and second, to spur the defense SOEs to act more like true industrial enterprises and therefore (1) be more responsive to their customer base (i.e., the PLA), and (2) reform, modernize, and "marketize" their business operations.

These goals in particular are central to the PLA's new requirements – as laid out in China's 2004 defense white paper – for fighting limited local wars under conditions of informatization.[30] This in turn is linked to a "generation leap" industrial strategy when it comes to armaments development and production – that is, skipping or shortening the stages of R&D and of generations of weapons systems. This process, according to You Ji, entails a "double construction" approach of mechanization and "informatization" in order to concurrently upgrade and digitize the PLA.[31] This "two-track" approach calls for both the near-term "upgrading of existing equipment combined with the selective introduction of new generations of conventional weapons" – a so-called "modernization-plus" approach – together with a longer-term "transformation" of the PLA along the lines of the information technologies-led "revolution in military affairs" (RMA).[32] Cheung argues that this plan was formalized in both the 2006–2020 Medium- and Long-Term Defense Science and Technology

Development Plan and the 11th Five-Year Defense Plan, both of which emphasized acceleration of PLA modernization and a new defense R&D drive.[33] Part of
this two-track approach also depends on China's "latecomer advantage" of being
able to more quickly exploit technological trails blazed by others, as well as being
to avoid their mistakes and blind alleys.[34]

The most recent and significant round of defense industry reforms began
around 1997, when the 15th Communist Party Congress laid out an ambitious agenda for restructuring and downsizing the state-owned enterprise sector (including the defense industries) and for opening up SOEs to free-market
forces – i.e., supply-and-demand dynamics, competitive products, quality assur-
ance, and fiscal self-responsibility. In March 1998 the 9th National People's
Congress (NPC) further refined this agenda by announcing plans to reorganize
the government's defense industry oversight, to control apparatus, and to establish new defense enterprise groups.

One of the most important decisions to come out of the 1998 NPC was the
demotion of COSTIND to an ancillary role in coordinating defense R&D and, at
the same time, the creation of a new PLA-run General Armaments Department
(GAD), which was intended to act as the primary purchasing agent for the PLA,
overseeing defense acquisition and new weapons programs. As a 2005 RAND
report put it, the GAD is part of a process "to create system that will unify, stand-
ardize, and legalize the [Chinese] weapons procurement process."[35] As such, the
GAD is supposed to ensure that local arms producers meet PLA requirements
when it comes to capabilities, quality, costs, and program milestones. The GAD
was given the authority to implement a "robust" regulatory, standards, and evaluation regime that would enforce quality control and performance and incentiv-
ize competition and innovation.[36]

More importantly, the establishment of the GAD exemplified a major change
in how the Chinese approached defense innovation. According to Cheung, the
Chinese military research, development, and acquisition (RDA) system has, since
the mid-1980s, gradually transitioned from a "technology-push" model (i.e.,
weapons programs driven mainly by what the defense industries can deliver) to
a "demand-pull" type – that is, driven by PLA requirements and "ensuring that
military end-user needs are being served."[37] This new RDA reform process was
only partly implemented with the creation of the GAD, which gave the PLA leading authority over defense innovation and R&D. In particular, the GAD had the
ability to concentrate R&D funding on "select high-priority projects," with the
intended effect of injecting a modicum of competition among R&D institutes
when it came to winning R&D work.[38]

Concurrently, COSTIND's role in overseeing the defense industrial base was substantially reduced, basically to "the making and administration of government policies towards the defense industry."[39] Oversight and administration of the defense
industry enterprises was placed under a new organization, the State-Owned Assets
Supervision and Administration Commission (SASAC), which reports directly to
the State Commission (the PRC's chief administrative authority). This diminished
status was followed by COSTIND's eventual demotion in 2008 from a ministerial

level entity to a bureau within the Ministry of Industry and Information Technology (MIIT), subsequently renamed the State Administration of Defense Science, Technology, and Industry (SASTIND).[40]

Another key element of current defense reforms was the creation in July 1999 of ten new "defense industry enterprise groups" (see Table 3.1). Basically, each existing SOE was divided into two new enterprise groups, e.g., AVIC was split into AVIC I and AVIC II, the China Aerospace Corporation was broken into China Aerospace Science & Technology Corporation (CASC) and the China Aerospace Science and Industry Corporation (CASIC), etc. Each new enterprise was given quite specific functions and responsibilities within the defense sector: AVIC I, for example, was to concentrate mainly on fighter aircraft, bombers, transports, and advanced trainer jets, while AVIC II had sole responsibility for helicopters, light trainer planes, and UAVs (AVIC I and II would later be reintegrated into a

Table 3.1 Restructuring the Chinese defense industry, 1999

Old Corporate Entity	New Enterprise Group	Major Products
Aviation Industries of China (AVIC)	China Aviation Industry Corp. I (AVIC I)	Fighter aircraft, bombers, transports, advanced trainers, commercial airliners
	China Aviation Industry Corp. II (AVIC II)	Helicopters, attack aircraft, light trainers, UAVs
China Aerospace Corp. (CASC)	China Aerospace Science & Technology Corp. (CASC)	Space-launch vehicles, satellites, missiles
	China Aerospace Science and Industry Corp. (CASIC)	Missiles, electronics, other equipment
China Ordnance Industry Corp. (COIC)/NORINCO	China North Industries Group Corp. (CNGC)	Tanks, armored vehicles, artillery, ordnance
	China South Industries Group Corp. (CSGC)	Miscellaneous ordnance, automobiles, motorcycles
China State Shipbuilding Corp. (CSSC)	China State Shipbuilding Corp. (CSSC)	Frigates, smaller surface combatants, commercial ships
	China Shipbuilding Industry Corp. (CSIC)	Destroyers, commercial ships
China National Nuclear Corp (CNNC)	China National Nuclear Corp. (CNNC)	Nuclear energy development, nuclear fuel and equipment
	China Nuclear Engineering & Construction Group Corp. (CNECC)	Construction of nuclear power plants, other heavy construction

single corporation in 2008). In 2002 an eleventh defense enterprise group, the China Electronics Technology Group Corporation (CETC), in charge of defense electronics R&D and production, was formed.

These enterprises were supposed to function as true conglomerates, integrating R&D, production, and marketing. In particular, by breaking up the old SOEs it was hoped that this would encourage the new industry enterprise groups to compete with each other for PLA procurement contracts, which in turn would cause them to be more efficient and technologically innovative. At the same time, the government's role in the daily operations of the defense industry was to be greatly reduced, and these new enterprise groups were given the authority to manage their own operations as well as to take responsibility for their own profits and losses. Defense enterprise groups were subsequently permitted to restructure their debt and to raise capital through limited public stock offerings and low-cost loans from state banks.[41]

Another crucial aspect of these new reform initiatives was the declared intent to significantly downsize the Chinese military-industrial complex, including eliminating (through retirement, attrition, or even layoffs) as much as one-third of the defense sector's workforce. The rationalization of the defense industry was also supposed to include factory closings and consolidation as a result of government-encouraged mergers, as part of the policy of "letting the strong annex the weak."[42]

To combat the endemic tendencies of a statist corporate culture – i.e., "a strong aversion to risk, a lack of competitive instincts, poor motivation, and weak disciplinary practices" –[43] COSTIND in the late 1990s launched the "Four Mechanisms" campaign. These mechanisms included *competition* (open up the arms production and procurement process to multiple actors), *evaluation* (improved auditing, costing, appraisal, and assessment of weapons projects), *supervision* (eliminate corruption on the part of management), and *encouragement* (foster a better work environment in order to develop a more motivated and innovative workforce).[44]

Finally, China began to seriously pursue the idea of leveraging advanced technologies and manufacturing processes found in the commercial sector in order to benefit defense R&D and production. According to many analysts, such civil-military integration (CMI) is a central feature of defense industry reform.[45] CMI is viewed as a fast (or at least faster) and ready means to shortcut the R&D process when it comes to advanced weapons systems; to cherry-pick civilian manufacturing practices in high-tech sectors (e.g., computer-aided design and manufacturing [CAD/CAM], program management tools, etc.); to exploit dual-use technologies (e.g., space systems for surveillance, communication, and navigation) to support the military; and, in particular, to take advantage of the latent capabilities found in commercially based information technologies (IT) in order harness the IT-based RMA. Such civil technologies could be both domestically developed or obtained from foreign sources via joint ventures, technology transfer, or even espionage.[46]

To a certain extent, CMI in the Chinese defense industry is nothing new. This strategy has its roots in the late 1970s and the enunciation of Deng Xiaoping's so-called 16-character slogan: "Combine the military and civil/combine peace and war/give priority to military products/let the civil support the military." However, whereas earlier efforts at CMI tended to revolve mostly around defense conversion – that is, switching military factories over to civilian use – China's approach to CMI after 1997 entailed a critical shift in policy toward promoting integrated dual-use industrial systems capable of developing and manufacturing both defense and military goods – or as one Western analyst put it, "swords into plowshares . . . and better swords."[47]

This new strategy is embodied in the principle of *Yujun Yumin* (locate military potential in civilian capabilities), enunciated at the 16th Party Congress in 2002.[48] Subsequently, *Yujun Yumin* has been made a priority in the last several Five-Year Defense Plans, as well as the 2006–2020 Medium- and Long-Term Defense Science and Technology Development Plan. These plans all emphasize the importance of the transfer of commercial technologies to military use, and they call upon the Chinese arms industry not only to develop dual-use technologies, but also to actively promote joint civil-military technology cooperation. Consequently, the spin-on of advanced commercial technologies, both to the Chinese military-industrial complex and in support of the overall modernization of the PLA, has been made explicit policy.[49]

The key areas of China's new focus on dual-use technology development and subsequent spin-on include microelectronics, space systems, new materials (such as composites and alloys), propulsion, missiles, computer-aided manufacturing, and particularly information technologies. Over the past decade Beijing has worked hard both to encourage further domestic development and growth in these sectors *and* to expand linkages and collaboration between China's military-industrial complex and civilian high-technology sectors. Factories were also encouraged to invest in new manufacturing technologies, such as CAD, computer numerically controlled (CNC) multi-axis machine tools, computer-integrated manufacturing systems (CIMS), and modular construction in shipbuilding, as well as to embrace Western management techniques. In 2002, for example, the Chinese government created a new industry enterprise group, the China Electronics Technology Group Corporation, to promote national technological and industrial developments in the area of defense-related electronics. Defense enterprises have formed partnerships with Chinese universities and civilian research institutes to establish technology incubators and to undertake cooperative R&D on dual-use technologies. Additionally, foreign high-tech firms wishing to invest in China have been pressured to set up joint R&D centers and to transfer more technology to China.

Assessing Chinese defense industrial base reforms

After decades of false starts and fitful progress, Beijing appears to have finally hit upon the right formula to reform and revitalize its defense industries, and since the turn of the century China has made significant progress in turning around its long-ailing defense sector. Beginning in the late 1990s and accelerating in the first decade of the 21st Century, Beijing launched several initiatives intended to inject more market-oriented thinking into the defense industry, including the introduction of more free-market practices, a new emphasis on quality control, and greater oversight by the Chinese military when it comes to procurement and program management. Efforts were also made to rationalize the country's bloated military-industrial complex, laying off excess workers and consolidating production. China even injected a modicum of competition, breaking up giant defense SOEs into smaller, contending firms, particularly in the aviation and ship-building sectors. At the same time, defense R&D was ramped up, and the production of several types of relatively advanced weapons systems began to emerge from Chinese factories and shipyards.

The proof that these post-1997 reforms have been at least partly successful is evident in the growing numbers of new types of weapons coming out of Chinese defense factories, armaments that are increasingly of a quality and capability comparable to Western systems. With regard to its naval forces, for example, China has built three different types of destroyers between 2000 and 2016, including one class (the Type-052C/D) which is outfitted with an *Aegis*-type air-defense radar and fire-control system; at least 18 Type-052C/D destroyers could eventually be constructed. These vessels are equipped with the indigenous YJ-83 or YJ-62 anti-ship cruise missile (ASCM) and the HN-2 land-attack cruise missile. The Type-052C/D also carries several Chinese-built HHQ-9 surface-to-air missiles (SAM), housed in vertical launch systems (VLS), for local air defense. A new, 14,000-ton Type-55 destroyer is also under construction, and it could be launched as early as 2018. The PLAN has also acquired nearly two dozen new stealthy Type-054A frigates (with more likely to be constructed), as well as 60-odd Type-022 *Houbei*-class catamaran-hulled missile fast attack craft.

China has also greatly expanded its submarine fleet over the past 15 years. Since the late 1990s the PLAN has acquired more than two dozen modern Type-039 *Song*-class and Type-41 *Yuan*-class diesel-electric submarines. These boats are the first Chinese-built submarines to feature a modern "Albacore" (or teardrop-shaped) hull and a skewed propeller (for improved quieting), and to carry an encapsulated ASCM capable of being fired while submerged (through a regular torpedo tube), as well as anti-submarine rockets. Additionally, there is speculation that the *Yuan*-class may be outfitted for air-independent propulsion (AIP) – perhaps the Stirling engine, which has been outfitted to Swedish and Japanese submarines.

As noted in the previous chapter, the PLAN has begun replacing its small and aging fleet of nuclear-powered submarines with two new classes of attack boats (SSN) and ballistic missile-carrying submarines (SSBN); these are the Type-093 *Shang*-class and the Type-94 *Jin*-class SSBN. The first Type-93 was launched in 2002 and at least five have been built. At least four new SSBNs of the Type-094 *Jin*-class, each carrying 12 7,000-kilomter-range JL-2 submarine-launched ballistic missiles (SLBMs), have also been launched, and four more are believed to be planned. Follow-on SSNs and SSBNs are already in development.[50]

Considerable attention has been paid as of late to the DF-21D anti-ship ballistic missile (ASBM). The first of its kind, the DF-21D ASBM combines a maneuverable reentry vehicle (MARV) with a terminal guidance system, has a range of 1,500 kilometers, and is capable of hypersonic (Mach 5 and above) speeds.[52] This makes the missile potentially effective against slow-moving carrier battle groups, which has earned the DF-21D the nickname "the carrier-killer." According to the US DoD, the DF-21D appears to be a "workable design" and has been deployed in small numbers, having achieved "initial operating capability."[53]

In perhaps its most dramatic development of late, the PLAN has recently taken delivery of China's first aircraft carrier, the rebuilt Soviet carrier ex-*Varyag*. Acquired from Ukraine in 2001, the *Varyag* was moved to a dry dock at the Dalian shipyard in northeast China in 2005, where it underwent extensive repairs and reconstruction, including the installation of new engines, radars, and electrical systems. The rebuilt ex-*Varyag* carrier underwent its first sea trials under PLAN colors in August 2011 and was re-commissioned the *Liaoning* in 2012. The *Liaoning* is currently undergoing takeoff and landing trials with the J-15 fighter jet (which is based on the Russian Su-33 acquired surreptitiously from Ukraine and subsequently reverse engineered). As of 2016, at least one, or perhaps even two, totally indigenous aircraft carriers are reportedly under construction at shipyards in Dalian and Shanghai.

Turning to the air forces, both the PLAAF and the PLAN Air Force (PLANAF) have, over the past 15 years, acquired a large number of so-called "fourth-generation" or "fourth-generation-plus" fighter aircraft. Beginning in 1992, for example, China began to import the Russian-built Su-27 *Flanker* fighter jet; this was quickly followed by an agreement to permit China to license-produce the Su-27 (designated the J-11A) at the Shenyang Aircraft Company. The Chinese then reverse engineered this aircraft, designated the J-11B, using all-Chinese avionics while still relying on a Russian-supplied engine.

China is also currently in full-scale production of its first indigenous fourth-generation-plus combat aircraft, the J-10. The J-10 is an agile fighter jet in roughly the same class as the F-16C, and it features fly-by-wire flight controls and a glass cockpit (but nevertheless equipped with the Russian AL-31 engine, underscoring China's continuing difficulties with developing a usable jet engine). While the J-10 had a rather rocky start, the program took on new momentum around the turn of the century, and perhaps 400 J-10s have been delivered to the PLAAF, with production continuing at a rate of about 30 aircraft a year.[54]

In addition, China has two "fifth-generation" combat aircraft programs – the J-20 and the J-31 – currently in the works. Each aircraft is the product of China's two leading fighter jet enterprises: the J-20 is being developed by the Chengdu Aircraft Industry Group, while the Shenyang Aircraft Corporation is in charge of the J-31 program. The J-20 first flew in January 2011, and the J-31 followed suit in October 2012. Both planes nominally resemble currently flying fifth-generation combat aircraft (that is, the US F-22 and F-35 Joint Strike Fighter), and in fact they may have benefited from industrial espionage aimed at these two US fighter programs.[55] At the same time, the actual details surrounding both aircraft – how stealthy they are, how advanced their radar and other avionics are, what kind of sophisticated weaponry do they carry, etc. – remains sketchy. Consequently, one should approach both programs cautiously and with skepticism as to whether they are truly fifth-generation fighters.[56] At the same time, the J-20 and J-31 programs demonstrate China's ambitions – and the aggressive steps it is prepared to take – to claw its way up into the vanguard of advanced fighter-jet producers. Consequently, China is currently producing three fourth-generation fighters – the J-10, J-11B, and J-15 – and it has two fifth-generation fighters in development.

The Chinese military has greatly upgraded its electronics, communications, and information systems over the past 15 years. According to Mulvenon and Tyroler-Cooper, the PLA is "in the midst of a C4I revolution, characterized by the wholesale shift to digital, secure communications via fiber optic cable, satellite, microwave, and encrypted high-frequency radio."[57] The PLA is also focused on developing its capacities for "integrated network electronic warfare," such as electronic defenses and countermeasures, computer network attacks, (that is, disrupting the enemy's computer networks), and physical attacks on the enemy's C4SIR networks, such as with antisatellite (ASAT) weapons.[58]

Finally, Chinese efforts at civil-military integration appear to be paying dividends. China's aggressive pursuit of advanced commercial technologies development and their subsequent spin-on onto the defense sector have shown some success in a number of areas, such as electronics and information technologies, shipbuilding, aviation, space-launch vehicles, satellites, and advanced manufacturing. In particular, China's military shipbuilding sector appears to have particularly benefited from CMI efforts over the past decade. Following an initial period of basically low-end commercial shipbuilding – such as bulk carriers and container ships – China's shipyards have since the mid-1990s progressed toward more sophisticated ship design and modular construction work. In particular, moving into commercial shipbuilding began to bear considerable fruit beginning in the late 1990s, as Chinese shipyards modernized and expanded operations, building huge new dry-docks, acquiring heavy-lift cranes and computerized cutting and welding tools, and more than doubling their shipbuilding capacity. At the same time, Chinese shipbuilders entered into a number of technical cooperation agreements and joint ventures with shipbuilding firms in Japan, South Korea, Germany, and other countries, which gave them access to advanced ship designs and manufacturing technologies – in particular, computer-assisted design and manufacturing, modular construction techniques, advanced ship propulsion systems, and numerically controlled processing and testing equipment. As a result, military shipbuilding programs – which are usually collocated at Chinese shipyards engaged in mostly commercial activities – has been able to leverage these considerable infrastructure and software improvements when it comes to design, development, and construction, and this is evident in the comparatively higher quality and capacity of warships being delivered to the PLAN.[59]

China's rapidly expanding space industry has also spurred the development and application of dual-use technologies that are basically commercial in nature but which serve military purposes as well. This includes telecommunications satellites, the *Beidou* navigation satellite system, and the *Yaogan* and *Ziyuan* types of earth observation satellites. In addition, many of the technologies being developed for commercial reconnaissance satellites, such as charge-coupled device cameras, multispectral scanners, and synthetic aperture radar imagers, have obvious spin-on potential for military systems.

Finally, the PLA has clearly profited from piggy-backing on the development and growth of the country's commercial information and communications technology (ICT) industry. As Mulvenon and Tyroler-Cooper point out, the Chinese military electronics, communications, and information systems has always been a special case when it comes to R&D and production, benefiting from a "digital triangle" comprised of the PLA (as a sponsor of commercial-to-military spin-on), China's increasingly sophisticated commercial ICT industry, and state laboratories, research institutes, and R&D funding institutions; in particular, the PLA, according to Mulvenon and Tyroler-Cooper, aided by the "growing use of COTS [commercial off-the-shelf technologies]," which permits it to "directly benefit from the globally competitive output of China's commercial IT companies."[60]

In addition to improving and expanding military R&D and production, and increasing the exploitation of commercial and dual-use technologies, the Chinese defense industrial base has undergone some successful structural reforms. For one thing, the defense industry appears to be serious when it comes to tackling problems of excess production capacity and workforce redundancy; according to Cheung, the Chinese defense industrial base shed nearly 1.5 million jobs between 1990 and 2006, declining from 3.15 million workers to 1.67 million.[61] Defense enterprise groups have also begun to find new sources of capital by selling shares in some their subsidiary firms on the Shanghai, Shenzhen, and Hong Kong stock markets; for instance, in 2007, Xian Aircraft International – an auxiliary company under the Xian Aircraft Industry Corporation, which manufacturers bombers and fighter-bombers for the PLAAF – was listed on the Shenzhen stock market for US$970 million.[62]

More importantly, the defense sector has erased much of the red ink on its balance sheets. In 2011, the ten leading state-owned defense enterprise groups posted combined profits of 80 billion yuan (US$12.1 billion); this compares with profits of 70 billion yuan (US$10.3 billion) earned in 2010, and with 43 billion yuan (US$5.5 billion) worth of profits in 2007.[63]

Consequently, China's defense industry appears to be better suited than ever to absorb and leverage advanced, militarily relevant technologies and therefore provide the PLA with the advanced military systems it requires. In fact, in recent years Beijing has greatly reduced its once-sizable arms purchases from Russia – according to the Stockholm International Peace Research Institute (SIPRI), Chinese arms imports fell from US$3.54 billion to 2005 to US$1.4 billion in 2014 –[64] an indicator that China is getting closer to realizing its long-cherished goal of self-sufficiency in arms acquisition.

All these accomplishments notwithstanding, a number of problems still confront the China's defense industrial base. For one thing, even after taking into account recent efforts at rationalization, the defense sector may still be over-capacitized for its needs and still too compartmentalized to be efficient. Cheung notes that the Chinese arms industry still suffers from the "widespread duplication and Balkanization of industrial and research facilities," of "around 1,400 large and medium-sized factories . . . scattered across the country, especially in its land-locked interior . . . possessing outdated manufacturing and research attributes." In addition, regional rivalries and "local protectionism" mean that there is "little cooperation or coordination among these facilities, preventing the exploitations of economies of scale and hampering efforts at consolidation."[65]

Consequently, there still exist critical shortcomings in the Chinese defense industry. In particular, it remains woefully deficient in the area of propulsion systems, including gas turbine engines for its destroyers, marine diesels for its diesel-electric submarines, and, above all, turbofan engines for its combat aircraft. Despite decades of effort, the country still lacks the ability to build a suitable jet aircraft engine, and it must continue to purchase engines from Russia to power its J-10 and J-11B fighters.[66] Foreign dependencies also still persist when it comes to turboshaft engines (for helicopters).

Structural reforms have also failed to produce much in the way of competition. Few arms programs are truly competitive; contracts are spread out among all the major defense firms (e.g., Chengdu builds the J-10, Shenyang produces the J-11B). Even more importantly, most defense enterprise groups are monopoly suppliers, without rivals. The China North Industries Group Corp. (CNGC), for instance, has sole responsibility for all tank, armored vehicle, and artillery production, while the two successors to the China Aerospace Corporation – China Aerospace Science & Technology Corp. (CASC) and China Aerospace Science and Industry Corp. (CASIC) – have practically no overlaps; CASC specializes in ballistic missiles, space-launch vehicles, and satellites, while CASIC manufactures mainly tactical missile systems, such as surface-to-air missiles, anti-ship cruise missiles, land-attack cruise missiles, and the like.[67] In fact, in 2008 the central authorities took the unusual step of re-merging AVIC I and II back into a single corporate entity, mostly because AVIC II was failing to secure lucrative contracts from China's growing commercial aircraft work.[68] About the only true competition in the Chinese arms industry that appears to exist can be found in the helicopter, trainer jet, and shipbuilding (destroyers and submarines) businesses.

The recent profitability of Chinese defense firms may also be called into question. While these enterprise groups may indeed be in the black, it is unclear how much this is due to purely military-related work and how much is due to commercial undertakings. Only a small portion of these 'defense conglomerates' revenues – around 10 percent to 25 percent of the value of total industrial output – is derived from defense work, and these companies provide no breakdown between profits from commercial work and those from defense contracting. According to Cheung, Chinese arms producers have "long complained" that they have to "struggle" to make money from defense work.[69]

Even China's apparent success with CMI may be less than meets the eye. It has been difficult for the central authorities to entice commercial enterprises to get into defense work or to partner with defense firms on joint projects that entail diffusing technologies and innovations to the military side. According to Cheung, less than one percent of China's commercial high-tech firms are engaged in defense work; consequently, "CMI has so far barely scratched the surface of the Chinese economy."[70] Impediments to deepening and broadening CMI include (1) weak institutions, mechanisms, and guidelines to promote and support CMI; (2) high barriers between civilian enterprises and the defense market; (3) corporate parochialism on both sides (commercial firms are often overly protective of their intellectual property, while military secrecy makes technology-sharing problematic); (4) insufficient resource-sharing; and (5) underdeveloped industries dedicated to CMI.[71] Overall, civilian firms are still only tangentially engaged in armaments production, and China has a long ways to go before it can claim widespread success in implementing CMI.

Lastly, China has experienced only modest success in overhauling the regulation and oversight of the defense industrial base. The GAD has limited authority when it comes to setting requirements and managing armaments programs. According to Cheung, the GAD only has responsibility for the ground forces, the People's Armed Police, and the militia, while the "navy, air force, and Second Artillery [nuclear weapons] have their own armaments bureaucracies," and competition between these various agencies for budgetary resources is fierce.[72] This limited impact is perhaps why the Chinese government decided to disband the GAD in early 2016 and to replace it a new Equipment Development Department under the direct authority of the Central Military Commission. For their part, the "four mechanisms" seem to have had little impact on reforming the corporate culture within the defense industry. Overall, therefore, China's defense industrial base remains a heavily statist affair, with armaments production still overwhelmingly situated in mostly noncompetitive and largely firewalled state-owned enterprises, protected and nurtured by the central government, with only modest changes to corporate management style (the major exception being the defense electronics industry, which is more closely tied to its commercial counterparts).

China's defense industrial base is certainly much more capable than it was in the late 1990s. The weapons systems coming out of its factories and shipyards are vastly superior to what was being produced less than 15 years ago. Progress in modernizing both armaments and armaments production is undeniable; at the same time, production facilities are humming, and the defense industry is turning unprecedented profits. What, then, are the keys to China's recent successes as a developer and manufacturer of advanced armaments?

Three factors have been perhaps the most critical in this regard: commitment, money, and technology acquisition. In the first place, China's military industrial complex has benefitted from an unwavering commitment on the part of national leadership (i.e., both the CCP and the PLA) to modernizing the country's defense R&D and manufacturing base. Reconstructing and upgrading the national defense industrial base has been a priority for more than 20 years. This commitment has in turn manifested itself in steady and large annual increases in Chinese military expenditures – meaning more money for innovation, more money for R&D, more money to increase procurement (and therefore production runs), and more money to upgrade the defense industrial base with new tools, new computers, and new technical skills. China has experienced double-digit real (i.e., after inflation) growth in defense spending nearly every year since the late 1990s. Even according to its own official national statistics, which most expert observers believe substantially understate spending levels, China's defense budget from 1999 to 2008 expanded at a rate of 16.2 percent *per annum*.[73] Most recently in March 2016, Beijing announced that it would allocate US$146 billion for defense, an increase of approximately 7.6 percent over 2015. Overall, since 1997 Chinese military expenditures have increased at least 600 percent in real terms. As a result, since the late 1990s China has moved up to become the second-largest defense spender in the world, outstripping Japan, France, Russia, and the United Kingdom; only the United States currently spends more on defense.

The impact of these increases in defense spending on defense R&D and procurement has been nothing less than astounding. In real terms PLA annual spending on equipment procurement has increased from around US$3.1 billion in 1997 to an estimated US$50 billion in 2015. Of this, perhaps US$10 billion is dedicated to defense R&D. This likely makes China the second-highest spender in the world in terms of procurement and perhaps the second- or third-highest when it comes to defense R&D spending.[74]

This upward trend is likely to continue for some time. In May 2006, for example, Beijing approved a 15-year national development plan for defense science and technology, with the goal of "transforming the PLA into a modernized, mechanized, IT-based force" by 2020.[75] This program is intended to boost military R&D spending, focusing on high-technology weapons systems (and specifically on "IT solutions"), supporting advanced manufacturing technologies, and cultivating collaborative international defense R&D efforts.[76]

Arguably, if anything has had a positive impact on the defense industry, it is this explosion in defense spending – by increasing procurement and therefore production, by expanding R&D spending, and by subsidizing the upgrading and modernization of arms-manufacturing facilities. Consequently, China's defense industrial base is better suited than ever to absorb and leverage advanced, militarily relevant technologies and therefore to provide the PLA with the advanced military systems it requires.

In addition to greater resources being made available to underwrite armaments production, the acquisition of new technologies – and especially foreign technologies – has had a significant effect on the growth and modernization of the Chinese military-industrial complex. China has undertaken several initiatives in recent decades to advance its military science and technology (S&T) base. These include the 863 Program (launched in 1986 to promote research in areas such as information technologies, spaceflight, lasers, new materials, biotechnology, and automation), the Torch Program (intended to commercialize new and advanced technologies, as well as to establish technology incubators and science parks), and, most recently, the 2006–2020 Medium- and Long-Term Defense Science and Technology Development Plan. Concurrently, it has greatly expanded its S&T education program, training a new generation of defense scientists, engineers, and technicians.[77]

While advancing indigenous defense S&T, China has embarked on an aggressive campaign to acquire and exploit foreign technologies. According to Hannas, Mulvenon, and Puglisi, this process of foreign technology acquisition is part of an unprecedented and aggressive effort directed by the central Chinese government – a "deliberate, state-sponsored project to circumvent the costs of research, overcome cultural disadvantages, and 'leapfrog' to the forefront by leveraging the creativity of other nations."[78] China, they assert, is engaged in a multipronged effort to gain foreign advanced technologies through both legal and illegal means. These include exploitation of open sources, technology transfer and joint research, the return of Western-trained Chinese students, and, of course, industrial espionage (both traditional and, increasingly, cyber-espionage). Hannas, Mulvenon, and Puglisi document a number of cases whereby Chinese intelligence organizations stole technology and other defense secrets from the West, and these were ostensibly incorporated (or will be incorporated) into Chinese weapons systems.[79]

Money and technology, of course, go hand in hand. Ian Anthony once stated that arms production is a "capital- and technology-intensive industry,"[80] and capital is a critical enabler of technology acquisition. Consequently, China's success as an emerging producer of advanced conventional weaponry – more than any structural, organizational, or cultural reform initiatives (or even greater efforts at civil-military integration) – is due mostly to a rather traditional, even prosaic strategy: throwing more money and technology at the problem of military modernization. It may be less glamorous than radical reform, but then again, one cannot argue with this approach's accomplishments.

At the same time, critical weaknesses remain. The Chinese arms industry still appears to possess only limited indigenous capabilities for cutting-edge defense R&D, and Western armaments producers continue to outpace China when it comes to most military technologies, particularly in areas such as propulsion and defense electronics. Overall, it is still more of a "fast follower," always playing technology "catch-up" or else being niche innovator when it comes to military R&D. Again, this is not necessarily a bad strategy to pursue. As Hannas, Mulvenon, and Puglisi put it:

> China's genius, as it were, is in putting together a system that capitalizes on
> its practical skill at adapting ideas to national projects, while compensating

for its inability to create those ideas by importing them quickly at little or no cost.[81]

Additionally, it may be acceptable to be niche innovator if one's military is only looking to gain asymmetric niche advantages, such as the PLA using an ASBM to attack aircraft carriers.

Overall, the Chinese defense industrial base has made undeniable advancements over the past decade and a half in terms of manufacturing new, relatively modern military systems, and so long as the defense budget continues to grow and the Chinese continue to be able to acquire and exploit foreign technologies, this pace of defense development and production will likely quicken in the decades ahead.

Notes

1 The Chinese defense industry has been the object of considerable study in recent years. See, for example, Richard A. Bitzinger, et al., "Locating China's Place in the Global Defense Economy," in Tai Ming Cheung, ed., *Forging China's Military Might: A New Framework for Assessing Innovation* (Baltimore, MD: Johns Hopkins University Press, 2013); Mikhail Barabanov, Vasiliy Kashin, and Konstantin Makienko, *Shooting Star: China's Military Machine in the 21st Century* (Minneapolis, MN: East View Press, 2012); Tai Ming Cheung, "The Chinese Defense Economy's Long March from Imitation to Innovation," *Journal of Strategic Studies*, Vol. 34, No. 3 (June 2011); James Mulvenon and Rebecca Samm Tyroler-Cooper, *China's Defense Industry on the Path of Reform*, prepared for the US-China Economic and Security Review Commission, October 2009; Tai Ming Cheung, "Dragon on the Horizon: China's Defense Industrial Renaissance," *Journal of Strategic Studies*, Vol. 32, No. 1 (February 2009); Tai Ming Cheung, *Fortifying China: The Struggle to Build a Modern Defense Economy* (Ithaca, NY: Cornell University Press, 2008); Richard A. Bitzinger, "Reforming China's Defense Industry: Progress in Spite of Itself?" *Korean Journal of Defense Analysis* (Fall 2007); Evan S. Medeiros, Roger Cliff, Keith Crane, and James C. Mulvenon, *A New Direction for China's Defense Industry* (Santa Monica, CA: RAND, 2005); Keith Crane, Roger Cliff, Evan S. Medeiros, James C. Mulvenon, and William H. Overholt, *Modernizing China's Military* (Santa Monica, CA: RAND, 2005); Evan S. Medeiros, *Analyzing China's Defense Industries and the Implications for Chinese Military Modernization* (Santa Monica, CA: RAND, February 2004); David Shambaugh, *Modernizing China's Military: Progress, Problems, and Prospects* (Berkeley: University of California Press, 2002), pp. 225–283; John Frankenstein, "China's Defense Industries: A New Course?" in James C. Mulvenon and Richard H. Yang, eds., *The People's Liberation Army in the Information Age* (Santa Monica, CA: RAND, 1999); Richard A. Bitzinger, "Going Places or Running in Place? China's Efforts to Leverage Advanced Technologies for Military Use," in Susan Puska, ed., *The PLA After Next* (Carlisle Barracks: SSI Press, 2000); John Frankenstein and Bates Gill, "Current and Future Challenges Facing Chinese Defense Industries," *China Quarterly* (June 1996).
2 Cheung, "The Chinese Defense Economy's Long March," p. 326.
3 Bates Gill and Taeho Kim, *China's Arms Acquisitions from Abroad: A Quest for "Superb and Secret Weapons"* (Stockholm: Stockholm International Peace Research Institute, 1995), pp. 2–3, 8–18.
4 Frankenstein, "China's Defense Industries: A New Course?" p. 190.

5 Frankenstein, "China's Defense Industries: A New Course?" pp. 191–192.

6 Medeiros, et al., *A New Direction for China's Defense Industry*, pp. 4–18.

7 Medeiros, et al., *A New Direction for China's Defense Industry*, pp. 161–162; Shambaugh, *Modernizing China's Military*, pp. 261–262.

8 Ken Allen, "PLAAF Modernization: An Assessment," in James Lilly and Chuck Downs, eds., *Crisis in the Taiwan Strait* (Washington, DC: NDU Press, 1997), p. 244.

9 *Jane's Fighting Ships, 1999–2000* (London: Jane's Information Group, 1999), pp. 119–120, 124–125.

10 Medeiros, et Al., *A New Direction for China's Defense Industry*, pp. 170–171.

11 Richard F. Grimmett, *Conventional Arms Transfers to Developing Nations, 1998–2005* (Washington, DC: US Congressional Research Service, October 23, 2006), pp. 56, 57.

12 *China Daily*, October 3, 1997.

13 Rao Gangcan, *Development and Outlook in Newbuilding Technology in China* (manuscript), 1998, p. 17.

14 Frankenstein, "China's Defense Industries: A New Course?" pp. 197–199; "Industry Embraces Market Forces," *Jane's Defense Weekly*, December 16, 1998, p. 28; Harlan Jencks, "COSTIND Is Dead, Long Live COSTIND! Restructuring China's Defense Scientific, Technical, and Industrial Sector," in James C. Mulvenon and Richard H. Yang, eds., *The People's Liberation Army in the Information Age* (Santa Monica, CA: RAND, 1999), p. 617.

15 Tai Ming Cheung, *Rejuvenating the Chinese Defense Economy: Present Developments and Future Trends*, Study of Innovation and Technology in China, Policy Brief No. 19, September 2011, p. 31.

16 Shambaugh, *Modernizing China's Military*, p. 277; Frankenstein and Gill, "Current and Future Challenges Facing Chinese Defense Industries," p. 403.

17 Allen, "PLAAF Modernization: An Assessment," p. 235.

18 Jonathan Pollack and James Mulvenon, *Assembled in China: Sino-U.S. Collaboration and the Chinese Civilian Aviation Industry*, ms. (Santa Monica, CA: RAND, August 1998), pp. 37, 47–48.

19 *Report of the Select Committee on U.S. National Security and Military/Commercial Concerns with the People's Republic of China*, Part 13 (Washington, DC: U.S. House of Representatives, 1999) Internet version (http://www.house.gov/coxreport/).

20 Stokes, *China's Strategic Modernization*, p. 136.

21 Dean Cheng, *Civil-Military Integration in the Chinese Aircraft Industry*, ms., April 1999, p. 6.

22 Cheng, *Civil-Military Integration in the Chinese Aircraft Industry*, p. 6.

23 Allen, "PLAAF Modernization: An Assessment," p. 235.

24 Pollack and Mulvenon, *Assembled in China*, pp. 37, 47–48.

25 Yuko Arayama and Panos Mourdoukoutas, *China Against Herself: Innovation or Imitation in Global Business?* Westport, CT: Quorum, 1999, pp. 69–82.

26 Yuko Arayama and Panos Mourdoukoutas, *China Against Herself: Innovation or Imitation in Global Business?* (Westport, CT: Quorum, 1999).

27 Pollack and Mulvenon, *Assembled in China*, pp. 45–46.

28 Pollack and Mulvenon, *Assembled in China*, p. 19.

29 Quoted in Larry M. Wortzel, *China's Military Potential* (Carlisle, PA: US Army War College, October 1998), p. 20.

30 See "Chapter 3: Revolution in Military Affairs with Chinese Characteristics," *China's National Defense in 2004* (Beijing: State Council Information Office, 2004) (http://www.fas.org/nuke/guide/china/doctrine/natdef2004.html).

31 You Ji, "China's Emerging National Strategy," *China Brief*, November 24, 2004.

32 Cheung, "Dragon on the Horizon," pp. 30–31.

33 Cheung, "Dragon on the Horizon," pp. 30–31. See also Richard A. Bitzinger, "Modernizing China's Military, 1997–2012," *China Perspectives*, No. 2011/11 (November 2011), pp. 7–8; and Richard A. Bitzinger, "China's 'Revolution in Military Affairs': How Fast? How Furious?" *Stockholm Journal of East Asian Studies* (December 2007).

34 Ji, "China's Emerging National Strategy."

35 Keith Crane, et al., *Modernizing China's Military*, p. 165.

36 Cheung, "Dragon on the Horizon," pp. 30, 36.

37 Tai Ming Cheung, *Innovation in China's Defense Research, Development, and Acquisition System*, Study of Innovation and Technology in China, Policy Brief No. 20, September 2011, pp. 35–36.

38 Cheung, *Innovation in China's Defense Research, Development, and Acquisition System*.

39 Cheung, "Dragon on the Horizon," p. 36.

40 Cheung, "Dragon on the Horizon," pp. 36–38; Barabanov, Kashin, and Makienko, *Shooting Star: China's Military Machine in the 21st Century*, pp. 3–4.

41 Cheung, "Dragon on the Horizon," pp. 40–41; Cheung, *Rejuvenating the Chinese Defense Economy*, p. 28; Cheung, "The Chinese Defense Economy's Long March from Imitation to Innovation," p. 339.

42 Cheung, "Dragon on the Horizon," p. 41; Cheung, "The Chinese Defense Economy's Long March from Imitation to Innovation," pp. 339–340.

43 Cheung, "The Chinese Defense Economy's Long March from Imitation to Innovation," p. 348.

44 Cheung, "Dragon on the Horizon," pp. 43–44.

45 Eric Hagt, "Emerging Grand Strategy for China's Defense Industry Reform," in Roy Kamphausen, David Lai, and Andrew Scobell, eds., *The PLA at Home and Abroad: Assessing the Operational Capabilities of China's Military* (Carlisle, PA: US Army War College, July 2010), pp. 481–484; Brian Lafferty, Aaron Shraberg, and Morgan Clemens, *China's Civil-Military Integration*, Study of Innovation and Technology in China (SITC), Research Brief 2013–10, January 2013, pp. 58; Mulvenon and Tyroler-Cooper, *China's Defense Industry on the Path of Reform*, pp. 57–58.

46 Hagt, "Emerging Grand Strategy for China's Defense Industry Reform," 514–518; Mulvenon and Tyroler-Cooper, *China's Defense Industry on the Path of Reform*, pp. 35–37, 38–43; Cheung, "Dragon on the Horizon," p. 47.

47 Paul H. Folta, *From Swords to Plowshares? Defense Industry Reform in the PRC* (Boulder, CO: Westview, 1992), p. 1.

48 Mulvenon and Tyroler-Cooper, *China's Defense Industry on the Path of Reform*, p. 5.

49 Hagt, "Emerging Grand Strategy for China's Defense Industry Reform," 481–484.

50 Ronald O'Rourke, "PLAN Force Structure: Submarines, Ships, and Aircraft" (paper presented to the CAPS-RAND-CEIP-NDU conference on The Chinese Navy: Expanding Capabilities, Evolving Roles?), pp. 4–9, 13–18; Sinodefense. com, "Type 094 (Jin Class) Nuclear Powered Missile Submarine" (http://www. sinodefence.com/navy/sub/type094jin.asp).

51 O'Rourke, *PLAN Force Structure*, p. 19; Sinodefense.com, "Type 071 Landing Platform Dock" (http://www.sinodefence.com/navy/amphibious/type071. asp).

52 Wendell Minnick, "China Developing Anti-Ship Ballistic Missiles," *Defense News*, January 14, 2008.

53 Tony Capaccio, "China Has 'Workable' Anti-Ship Missile Design, Pentagon Says," *Bloomberg*, August 26, 2011 (http://www.bloomberg.com/news/2011–08–25/china-has-workable-anti-ship-missile-design-pentagon-says.html).

54 "AMR Regional Air Force Directory 2012," *Asian Military Review*, August 30, 2012 (http://www.asianmilitaryreview.com/directories).

55 "Pentagon Aircraft, Missile Defense Programs said Target of China Cyber Threat," *Washington Post*, May 29, 2013.

56 "J-20 vs. F-35, one analyst's perspective," *Defensetech.org*, December 31, 2010 (http://defensetech.org/2010/12/31/j-20-vs-f-35-one-analysts-perspective).

57 Mulvenon and Tyroler-Cooper, *China's Defense Industry on the Path of Reform*, p. 35.

58 US Department of Defense, *Annual Report on the Military Power of the People's Republic of China 2009* (Washington, DC: US Department of Defense, 2009), pp. 25–28; You Ji, "China's Emerging National Defense Strategy," Association for Asian Research, January 12, 2005; Wendell Minnick, "China Shifts Spending Focus to Info War," *Defense News*, September 11, 2006; and Bill Gertz, "Inside the Ring: China Info Warfare," *Washington Times*, June 2, 2010; Richard Bitzinger, "China's RMA: Good enough?" *International Relations and Security Network*, August 29, 2007.

59 Medeiros, et al., *A New Direction for China's Defense Industry*, pp. 140–152.

60 Mulvenon and Tyroler-Cooper, *China's Defense Industry on the Path of Reform*, pp. 35–37.

61 Tai Ming Cheung, "Charting the Chinese Defense Economy" (powerpoint briefing), slide no. 6, in Tai Ming Cheung, ed., *New Perspectives on Assessing the Chinese Defense Economy* (La Jolla, California: University of California Institute on Global Conflict and Cooperation, 2011).

62 Cheung, "Dragon on the Horizon," pp. 42–43.

63 Tai Ming Cheung, *The Chinese Defense Economy in the Early 2010s*, Study of Innovation and Technology in China (SITC), Research Brief 2013–1, January 2013, p. 18; Cheung, "Dragon on the Horizon," p. 41.

64 Stockholm International Peace Research Institute, "The SIPRI Arms Transfers Database" (http://armstrade.sipri.org/armstrade/html/export_values.php).

65 Cheung, *Rejuvenating the Chinese Defense Economy*, p. 31.

66 Barabanov, Kashin, and Makienko, *Shooting Star: China's Military Machine in the 21st Century*, pp. 35–48.

67 Barabanov, Kashin, and Makienko, *Shooting Star: China's Military Machine in the 21st Century*, pp. 13–15.

68 China currently has two large commercial airliners in the works, the ARJ-21 regional jet and the C919; however, only companies in AVIC I were chosen to work on these projects. By re-merging the authorities hope that subcontracts for these civilian programs can be spread to former AVIC II businesses. See Richard A. Bitzinger, "China and Commercial Aircraft Production: Harder Than It Looks," *China Brief*, Vol. 13, No. 2 (January 18, 2013).

69 Cheung, *The Chinese Defense Economy in the Early 2010s*, pp. 18–19.

70 Cheung, "The Chinese Defense Economy's Long March from Imitation to Innovation," pp. 343–344.

71 Lafferty, Shraberg, and Clemens, *China's Civil-Military Integration*, pp. 58–60.

72 Cheung, *The Chinese Defense Economy in the Early 2010s*, p. 20.

73 "China Plans to Boost 2009 Military Spending by 14.9%," *Bloomberg*, March 4, 2009.

74 In 2009, for example, France spent approximately US$14 billion on procurement and US$5.8 billion on defense R&D; that same year, the United Kingdom spent US$10.9 billion and US$4.2 billion, respectively, on procurement and R&D. EDA, *Defense Data of EDA Participating Member States in 2009*, p. 11. China defense R&D budget is unknown, but based on its overall military spending,

it is not unreasonable to assume that the Chinese allocate anywhere between US$5 billion and US$10 billion on military R&D.

75 Ben Vogel, "China Embarks on 15-Year Armed Forces Modernization Program," *Jane's Defense Weekly*, July 1, 2006.

76 Vogel, "China Embarks on 15-Year Armed Forces Modernization Program"; OSD, *2011 Report to Congress*, p. 45.

77 Cheung, "Dragon on the Horizon," pp. 52–54.

78 William C. Hannas, James Mulvenon, and Anna B. Puglisi, *Chinese Industrial Espionage: Technology Acquisition and Military Modernization* (New York: Routledge, 2013), p. 78.

79 See Hannas, Mulvenon, and Puglisi, "Appendix I," *Chinese Industrial Espionage*, pp. 256–270.

80 Ian Anthony, "The 'Third Tier' Countries: Production of Major Weapons," in Herbert Wulf, ed., *Arms Industry Limited* (Oxford: Oxford University Press, 1993), p. 365.

81 Hannas, Mulvenon, and Puglisi, *Chinese Industrial Espionage*, pp. 241.

4 India

India, like China, is an aspiring great power that has long harbored the goal of possessing a technologically advanced self-sufficient arms industry.[1] These ambitions go back more than 50 years, when the country attempted to design and build its own fighter aircraft, the HF-24 *Marut*. Although a technological failure, it did not dampen India's determination to one day becoming a major arms-producing nation, capable of meeting most, if not all, of its requirements for self-defense – and therefore great-power status – through indigenous means. This quest for autarky *and* stature, for example, drove the country's nuclear weapons program. As India's economic power has expanded, and as its technological prowess in certain areas (such as information technologies) has grown, it has become more determined than ever to create a world-class, globally competitive defense industry.

India possesses one of the largest and most broad-based defense industries in the developing world. It produces fighter aircraft, surface combatants, submarines, tanks, armored vehicles, helicopters, artillery systems, and small arms. The country also has a huge defense research and development (R&D) establishment with considerable experience in indigenous weapons design and development, going back more than 50 years. That said, India has long been confronted with serious impediments to its efforts to build a state-of-the-art arms industry. While the rest of India appears to be racing into the 21st century, powered by a dynamic, free market-oriented economy, the defense sector seems mired in the country's Nehruvian socialist and protectionist past. Consequently, the nation is still predominantly saddled with a bloated, non-competitive, non-responsive military-industrial complex – capable, it seems, of only producing technologically inferior military equipment, and even then never on time and nearly always way over their original cost estimates. Given such longstanding deficiencies in its defense industrial base, it is little wonder why India's drive for great power status has been so fitful.

This may be changing. The economic liberalization that began in India 20 years ago may finally be pervading the local arms industry. For more than a decade the Indian government has been engaged in a number of initiatives designed to open up the defense sector to competition; more recently, too, it has expanded efforts to bring in foreign technologies to improve the capabilities of home-grown armaments and to establish the foundation for a more high-tech defense R&D base. That said, many of these reforms continue to face stiff resistance, and for the present it is still uncertain what impact, if any, these efforts may eventually have on invigorating the Indian military-industrial complex.

India's traditional policy of self-reliance in arms production

Self-reliance has long been a fundamental goal of indigenous armaments production in India – such an objective had military, political, and economic salience. As Ajay Singh puts it:

> After independence, and the adoption of a policy of non-alignment, it was . . . obvious that foreign policy would need to be reinforced by a policy of self-reliance in defense . . . Prime Minister Jawaharlal Nehru believed that no country was truly independent, unless it was independent in matters of armaments.[2]

Early on, too, a distinction was made between "self-sufficiency" and "self-reliance." Singh defined the former as requiring that "all stages in defense production (starting from design to manufacture, including raw materials) . . . be carried out within the country." He added, "To be self-sufficient, a country must not only have the material resources required for defense production, but also the technical expertise to undertake design and development without external assistance." Self-*reliance*, on the other hand, was much more modest, as while it entailed the indigenous production of armaments, it allowed for the importation of foreign designs, technologies, systems, and manufacturing know-how.[3]

While self-sufficiency was the preferred approach, self-reliance has long been the practice when it comes to Indian armaments production. As such, New Delhi has long conceded the need to import considerable amounts of foreign military technology – mostly from the Soviet Union/Russia but also from France and the United Kingdom – in order to establish and expand its indigenous military-industrial complex. Thus, from the early 1960s to the late 1980s India undertook the licensed-production of several foreign weapons systems, including MiG-21 and MiG-27 fighter jets, Jaguar strike aircraft, Alouette III helicopters, T-55 and T-72 tanks, Milan antitank weapons, and *Tarantul* corvettes.[4]

At the same time, however, it was always New Delhi's intention to gradually and incrementally replace licensed-production with indigenously developed and designed weaponry. Consequently, starting as far back as the 1950s the manufacture of foreign-sourced military systems was complemented with local products.[5] India began development of its first indigenous fighter jet, the HF-24 *Marut*, in 1956, with the first flight occurring in 1961. Truly indigenous armaments development and production, however, did not really take off until the 1980s with the inauguration of several ambitious home-grown projects, such as the Light Combat Aircraft (LCA, renamed the *Tejas* in 2005), the Advanced Light

Helicopter (ALH), the *Arjun* tank, and, especially, the Integrated Guided Missile Development Program (IGMDP), which involved the development of a number of tactical missile systems. While many of these "indigenous" programs still incorporated considerable amounts of foreign technology or subsystems, the objective has always been to reduce this dependency along the lines of the evolutionary "ladder-of-production" model, and eventually to achieve true "self-sufficiency."[6] This intent was underscored, for example, in 1995 when New Delhi announced that within 10 years it would increase its "local content" of weapons systems in the Indian armed forces from 30 percent to 70 percent.[7]

India's military-industrial complex: an overview

Traditionally, Indian armaments production has been entirely embedded within a huge government-run military-industrial complex. Even in the second decade of the 21st century, after a few modest reforms (discussed later on), the vast bulk of defense manufacturing remains in the hands of the state. As such, the Indian defense industrial base consists of 8 government-owned Defense Public Sector Undertakings (DPSUs), 41 Ordnance Factories (OFs), and, at the top, the powerful Defense Research and Development Organization (DRDO).[8] The Indian state-run defense sector employs more than 1.4 million workers (of which about 105,000 work in the OFs), including some 30,000 scientists and engineers within the DRDO, and in 2010 it enjoyed revenues of approximately US$7.8 billion.[9]

The DPSUs and OFs carry out the bulk of Indian arms manufacturing, often operating mainly as monopoly suppliers. Hindustan Aeronautics Ltd. (HAL), for example, is the sole DPSU engaged in aircraft production, including combat aircraft, helicopters, trainers, and transport planes, as well as avionics and engines.[10] HAL was established in 1964 with the merger of Hindustan Aircraft Ltd. and Aeronautics India Ltd.; it is headquartered in Bangalore and operates four main manufacturing and design complexes. HAL both license-produces foreign-designed aircraft – including the Soviet/Russian MiG-21, MiG-27, MiG-29, and (currently) Su-30 fighter jets, as well as the Anglo-French *Jaguar* strike aircraft – and manufactures indigenously developed combat planes, such as the HF-24 *Marut* and, currently, the *Tejas* LCA. Other military aircraft programs include the Advanced Light Helicopter (ALH), the Light Combat Helicopter (LCH), and the Intermediate Jet Trainer (IJT).

Bharat Dynamics Ltd. (BDL) builds tactical and strategic missiles for the Indian military. Most important, BDL is the production base for India's Integrated Guided Missile Development Program (IGMDP), which was launched in the early 1980s. The IGMDP entailed the development and production of several types of missile systems: initially two surface-to-surface ballistic missiles (the short-range *Prithvi* and the medium-range *Agni*), the *Akash* and *Trishul* surface-to-air missiles, and the *Nag* antitank guided missile (ATGM). Additionally, Bharat Dynamics builds the *Brahmos* anti-ship cruise missile, the *Sagarika* submarine-launched ballistic missile, and the *Astra* air-to-air missile.

Bharat Electronics Ltd. (BEL) is India's DPSU responsible for the production of electronic systems for the Indian armed forces. Established in 1954, BEL is based in Bangalore and operates nine factories producing radios and other communication gear, radars, sonars, electronic warfare systems, opto-electronics, and electronic components for tanks and other weapons systems.[11] BEL is also designing and developing the Indian Army's Tactical Communications System, as well as the Battle Management System, an army-wide network-centric "situational awareness" solution linking and integrating data from a wide range of sensors and transmitting this information in near real time to forces on the battlefield.[12]

The various Ordnance Factories are responsible for the manufacture of ground forces and miscellaneous military equipment, such as tanks and armored vehicles, artillery systems, small arms and ammunition, uniforms, tents, etc. The OFs are directly operated under the auspices of the Indian Ministry of Defense and administered through the Ordnance Factory Board (OFB). For several years the government operated 39 OFs, split into 5 operating divisions: Ammunition and Explosives (10 factories); Weapons, Vehicles, and Equipment (10 factories); Materials and Components (9 factories); Armored Vehicles (5 factories); and the "Ordnance Equipment Group of Factories" (5 factories). One of the more important OFs was the Heavy Vehicles Factory (HVF) near Chennai, which manufactures all main battle tanks for the India Army; currently, the HVF produces the indigenous *Arjun* tank and license-produces the Russian T-90. In 2010, 2 more OFs were established, raising the total number of OFs to 41.[13]

Intra-sectoral competition appears to exist only in the shipbuilding industry. The three chief DPSUs in charge of naval construction are Mazagon Dock Ltd. (MDL), Garden Reach Shipbuilders and Engineers Ltd., and Goa Shipyard Ltd. Mazagon Dock, located in Mumbai, is the country's oldest shipyard, founded in 1934 and nationalized in 1960. MDL is India's main naval shipbuilder; in the past it has produced the *Delhi*-class destroyers, the *Godavari*-class frigate, and the *Khukri*-class corvette, as well as German Type-209 submarines assembled under license. Currently, it is building the *Kolkata*-class destroyer, the *Shivailk*-class frigate, and the Franco-Spanish *Scorpène*-class submarine (six of which are currently being constructed under license). Garden Reach Shipbuilders and Engineers Ltd. is based in Kolkata and was founded in 1960. It is currently building the *Kamorta*-class corvette, along with various fast-attack craft and patrol vessels. Goa Shipyard Ltd., founded in 1967 and based in Vasco Da Gama (Goa), produces offshore patrol vessels, missile corvettes, and fast patrol vessels.

Interestingly, India's indigenous aircraft carrier, the INS *Vikrant*, which is currently under construction, is *not* being built at a naval DPSU, but rather at the Cochin Shipyard in Kochi. Cochin was traditionally a commercial shipbuilder (but still a state-owned firm, or PSU) manufacturing bulk carriers, tankers, and platform supply vessels. Given the likely construction of two or more indigenous carriers, however, Cochin could come to compete heavily with other shipbuilding DPSUs for naval contracts.

At the very top of India's military-industrial complex stands the DRDO. The Defense Research and Development Organization has primary responsibility for the design, manufacture, and management of indigenous weapons programs for the Indian armed forces. The DRDO comprises more than 50 state-owned laboratories engaged in the research and development of defense technologies; it employs over 30,000 workers, including 5,000 scientists and about 25,000 other scientific, technical, and supporting personnel. The DRDO's budget in 2010 was approximately US$1.88 billion, or 6 percent of overall Indian military expenditures.[14]

The DRDO is presently engaged in over 400 research projects, such as the development of missile systems, combat and trainer aircraft, radars, electronic warfare systems, and other types of armaments. Key R&D programs include the *Tejas* LCA, the next-generation Advanced Medium Combat Aircraft (AMCA), an advanced unmanned aerial vehicle, an airborne warning and control system for the Indian Air Force, and a "mini nuclear submarine" for the Indian Navy.[15] In addition, the organization has primary responsibility for all indigenous missile development programs, particularly the *Brahmos* anti-ship cruise missile, the *Shaurya* and *Sagarika* sea-based missiles, and the entire Integrated Guided Missile Development Program. The DRDO also manages the Aeronautical Development Agency (ADA), a consortium of over 100 defense labs and academic and industrial institutions established in the mid-1980s to specifically oversee R&D of all aspects of the *Tejas* Light Combat Aircraft, including airframe, propulsion, radar, and flight control systems.[16]

The DRDO has traditionally had very close ties to the DPSUs and OFs. In particular, the DRDO has acted as the Defense Ministry's principle investigator and evaluator of defense procurement programs. Consequently, the organization frequently serves as the mediator between the military services and the local defense industry, particularly when it comes to determining requirements and coordinating weapons, R&D, and production.[17]

To pay for all this, India has greatly increased military expenditures in recent years. Indian defense spending grew by two-thirds between 1998 and 2008, according to data provided by the Stockholm International Peace Research Institute (SIPRI).[18] In 2011 the Indian defense budget stood at US$36.5 billion, a rise of 11.6 percent over the previous year; this was equal to 1.83 percent of the country's gross domestic product (GDP). Procurement alone grew by 14 percent in 2011 to US$15.4 billion; of this amount, US$6.6 billion went to the Indian Air Force (IAF), US$4.2 billion to the army, and US$2.9 billion to the navy. Approximately US$1 billion was allocated for defense R&D.[19] By 2015 Indian defense spending totaled US$40 billion (however, still less than one-third of China's defense spending).

Enduring and endemic problems in India's defense industry

Despite more than 50 years of effort, the history of India's defense industry is a nearly unbroken story of spectacular failures. For several decades the Indian armaments production process has been a vicious cycle of ambitious program overreach followed by technological setbacks and lengthy delays, too often result-ing in equipment that has typically been of substandard quality and suboptimal performance. In 2006, for example, a government audit of the Ordnance Facto-ries revealed that about 40 percent of OF products had "not achieved the desired level of quality despite the fact that most items were in production for decades." This included T-72 tanks built under license, the INSAS assault rifle, and various ammunition.[20] Overall, the technology gap between Indian and foreign weapons systems has widened over the past two decades as the country has tried, unsuc-cessfully in most cases, to move from self-reliance to self-sufficiency.[21] At the same time costs have skyrocketed; according to one source, the country's five most important weapons programs – including the *Tejas* fighter, the *Arjun* tank, and the *Kaveri* engine – are at least two and a half times over their original budgets.[22]

For example, India's supposedly state-of-the art *Tejas* fighter jet is more than twelve years behind schedule, while R&D costs have nearly doubled. The *Tejas* LCA was intended to propel India's aerospace industry into the 21st century, advancing this sector in several key areas, including composites (carbon-fiber composites account for 45 percent of the aircraft, by weight), a modern "glass cockpit," a fly-by-wire (FBW) flight-control system, a multimode pulse-Doppler radar, and an afterburning turbofan engine. Unfortunately, India has run into several development problems regarding the *Tejas*, including delays in finalizing the aircraft's FBW software, and, more significantly, failures to deliver both an effective indigenous radar *and* jet engine; in both cases, a foreign substitute had to be found.[23] In particular, the indigenous *Kaveri* engine has suffered so many setbacks (in particular, it has been deemed overweight and underpowered) that it was "de-linked" from the *Tejas* program in 2008, and for the foreseeable future all *Tejas* aircraft will be outfitted with the US General Electric F404 turbofan.

Originally, the *Tejas* was to be initially deployed with the Indian Air Force around 2002, but the first "proof-of-concept" model did not fly until 2001, and a production-model LCA did not achieve first flight until 2007; the aircraft finally achieved initial operating capability with the IAF in 2011. Up to 260 LCAs could be built, in both IAF and naval versions (for India's future aircraft carriers), but so far only 40 aircraft have been ordered. The *Tejas* went into production in 2010 and will be manufactured as a very low rate of around 10–12 aircraft a year for the next 20 years; at that rate the aircraft could be obsolete before the last one is delivered.

For its part, the *Arjun* main battle tank only just entered service with the Indian Army (IA) in 2011, more than 30 years after the program was initiated. The *Arjun* has had a history of technical problems, resulting in horrendous delays and cost overruns – according to some sources, the tank is more than 16 years behind schedule and 20 times over its original cost estimates.[24] Its engine had a tendency to overheat, while its excessive weight and width made it too big for the IA's current tank transporters. The interior would become so hot that the tank's fire control system, thermal imager, and laser range-finder would be rendered useless; its rifled gun barrel prevents it from firing anti-tank rockets. In addi-tion, government reports have pointed to problems with the tank's powerpacks, "low accuracy and consistency," "failure of hydropneumatic suspension," and "chipping of gun barrels."[25] So far, the army has committed to buying only 248 *Arjuns*.

Even the country's much-vaunted Integrated Guided Missile Development Program, initiated in 1983 as a comprehensive, intensive effort to make India self-sufficient in tactical missile systems, has so far produced few successes. Only two IGMDP projects – the *Prithvi* and *Agni* surface-to-surface ballistic missiles – have so far been deployed to the Indian armed forces. Even then, the *Prithvi* is a relatively short-range, liquid-fueled missile (maximum range: 350 kilometers) of limited tactical use, while the *Agni* "does not appear to have been produced in large enough numbers for induction into the services."[26] Moreover, some missiles under the IGMDP, including the *Trishul* surface-to-air missile system and the *Astra* air-to-air missile, are still in development decades later and will likely never be anything more than "technology demonstrators."[27] For its part, the *Nag* ATGM is still undergoing test and validation trials after more than 20 years of development, and it was only accepted into the Indian Army (on a trial basis, with just 443 missiles being ordered) in 2011.[28] So far, only the *Akash* surface-to-air missile has gone into serial production and deployment.

The Indian defense industry even apparently has problems with programs as simple and technologically straightforward as small arms. The INSAS, the Indian Army's standard assault rifle, costs nearly US$400 apiece, or three times that of an imported AK-47.[29] Even so, the INSAS was found to have a number of defects, including poor performance in extremely cold and high-altitude situations; for these reasons it was removed from use in the Siachen glacier area. The Nepalese Army, which bought 100,000 INSAS rifles, claimed that the gun could only operate for about an hour or two before malfunctioning, resulting in heavy casualties in firefights with Maoist guerrillas.[30]

Overall, endemic delays and setbacks in domestic weapons programs have forced the Indian military to continually scrounge for foreign stopgaps to compensate for these shortfalls and to sustain force recapitalization. For example, due to ongoing delays in the *Tejas* program the Indian Air Force in the mid-2000s instituted the Medium Multi-Role Combat Aircraft (MMRCA) competition to buy 126 foreign fighter jets (with an option for up to 74 additional aircraft), at a cost of up to US$10 billion; in 2012 the IAF, after assessing six different fighters from Britain, France, Russia, Sweden, and the United States, eventually chose the French *Rafale*.[31] In addition, the IAF is currently acquiring up to 240 Russian Su-30MKIs, which are being licensed-produced by HAL. For its part, because of interruptions in the indigenous *Arjun* program, the Indian Army is buying several hundred Russian T-90 tanks, again locally built at Indian ordnance factories; the IA could acquire up to 1,400 T-90s, which would severely cut into purchases of the *Arjun*.[32] The IA is also buying 15,000 Russian-made *Konkurs*-M and 4,100 French *Milan*-2T antitank missiles due to setbacks in the *Nag* program; both will be licensed-assembled by Bharat Dynamics.[33] Finally, the Indian Navy has had to acquire Russian and Israeli surface-to-air missiles for its ships because local missile systems are still unavailable.

If anything, Indian efforts to move from self-reliance to self-sufficiency in armaments production has taken a huge step backwards over the past 15 years. The Indian military is as dependent as ever on foreign systems and technologies. Around 60 percent of the components for the *Arjun* tank are imported, for example, while the *Tejas* fighter utilizes a US jet engine and either a European or an Israeli radar.[34] Even India's highly touted *Brahmos* supersonic cruise missile (available in both anti-ship and land-attack variants) is heavily based on the Russian-designed P-800 *Yakhont* missile; India's particular contribution to this program, other than money, is hard to identify.[35] Indeed, the most advanced armaments coming out of Indian factories are still predominantly licensed-produced versions of foreign weapons systems: the Su-30MKI combat aircraft, T-90 tanks, the *Konkurs* and *Milan* ATGMs, *Scorpène* submarines, etc. Additionally, most of the aircraft to be acquired under the MMRCA program were supposed to be built indigenously, under license.[36]

Consequently, the Indian arms industry still functions mainly as an assembler rather than as an across-the-board innovator.[37] In 1995 New Delhi announced that within 10 years it would increase its "local content" of weapons in the Indian armed forces from 30 percent to 70 percent. By 2005, however, foreign weapons systems (that is, both imports and licensed-production) still comprised around 70 percent of the Indian military's acquisitions.[38] In 2010 the percentage of imported systems still hovered around 70 percent.[39]

At the same time India has become the largest arms buyer in the world; according to SIPRI, during the period of 2010–2014 New Delhi imported US$21 billion worth of arms, accounting for 15 percent of global arms deliveries. In addition to the aforementioned systems that will be licensed-built in India, the country has made off-the-shelf purchases such as the Phalcon airborne early-warning aircraft, Barak surface-to-air missiles, and UAVs from Israel; C-130J and C-17 transport aircraft, P-8 maritime patrol aircraft, and artillery-locating radar from the United States; and lightweight howitzers from the United Kingdom.[40]

The problems with India's defense industry are structural, financial, and, most of all, cultural. A cabal of monopolistic state-owned enterprises has traditionally dominated the arms-production process. In turn, these DPSUs and OFs are larded with bloated workforces and excess productive capacity; estimates are that much of the defense industry operates at barely 50 percent of capacity.[41] At the same time, the defense industry has been starved of capital for modernization and for keeping pace with the global state of the art in arms production. India's defense budget constituted less than 2 percent of the country's GDP in 2011.[42] Funding for defense R&D amounted to barely US$1 billion in 2011, barely 3 percent of total military expenditures; in contrast, the United States spent US$78 billion on defense R&D in FY2010, while China's military R&D budget is estimated to be around US$5 billion to US$6 billion. One result has been that the Indian defense sector has been unable to train enough highly qualified technicians, engineers, and scientists.[43] Finally, there has also traditionally been a lack of coordination between the defense sector and the armed forces when it comes to requirements, planning, and production.[44]

Despite these obvious deficiencies, there was for a long time little incentive from within the arms industry to reform and restructure itself. A "statist" mindset generally permeated the Indian military-industrial complex, and the government, DPSUs, and OFs operated in a cozy, sealed environment. Under the guise of "self-reliance," state-run defense firms were pretty much guaranteed production work; little stress was put on meeting project milestones or ensuring quality or operational effectiveness. Moreover, the private sector was not permitted to bid on major weapons contracts. For their part, the Indian armed forces were essentially forced to accept indigenous military equipment, whatever their preferences.[45] Consequently, as recently as 2005 one Indian defense ministry official was quoted as stating that "the DPSUs have no need to be competitive as they face no competition and have a captive market in the military."[46]

At the same time, defense industry employees were organized within a powerful union, and altogether these workers constituted an influential "vote bank." This in turn made it difficult to shed excess labor or to engage in other kinds of structural reforms, such as privatization or plant closures. Even where some downsizing was achieved – the Ordnance Factories, for example, cut their workforce from 150,000 in 1989 to 105,000 in 2010 – this was accomplished mainly by instituting a hiring freeze, resulting in the loss of new talent; moreover, according to Rahul Bedi, personnel reductions in many OFs were "lopsided," resulting in labor surpluses in those factories "where production lines face closure or are winding down."[47]

Much of the blame for the failure of the Indian military-industrial complex to perform adequately has been laid directly upon the Defense Research and Development Organization. The DRDO has frequently been criticized for its poor performance in overseeing the country's overall weapons development process.[48] In particular, the institution has been accused of arrogance, self-promotion, and weak leadership, with a stronger emphasis on the *acquisition* of technology and know-how than on its actual *application*. Cohen and Dasgupta, in their 2010 book on Indian military modernization, put it bluntly:

> The reasons for DRDO's failures are multifaceted. One review concluded that poor planning, over-optimistic timelines, and a lack of coordination with the armed forces led to cost and time overruns of major defense projects. However, the most important reason is the agency's lack of political leadership. DRDO officials engaged in exaggerated and wildly over-optimistic statements of their own capabilities, and civilian politicians with little knowledge about strategic or military affairs, alone the intricacies of military technology and hardware, allowed DRDO a free hand for decades.[49]

Insisting that maintaining an indigenous defense R&D and industrial base is a strategic technological and economic imperative, the DRDO historically took a reflexive approach that overwhelmingly and relentlessly favored indigenous solutions over foreign options. Particularly during the 1980s and 1990s, when India began its attempts to move from licensed-production-based "self-reliance" to more autarkic "self-sufficiency," the DRDO

> made it a practice to claim that it could provide services in, and make any product related to, aeronautics, armaments, electronics, combat vehicles, engineering systems, instrumentation, missiles, advanced computing and simulation, special materials, naval systems, life sciences, training and information systems.[50]

Consequently, the organization has had the persistent tendency to overestimate the technological abilities of the local defense sector while also lowballing weapons costs and development timelines:

> The organization has adopted a classic foot-in-the-door strategy: winning initial support by promising products on the cheap but later citing sunk costs to demand more money.[51]

At the same time, the DRDO has long had "the power to kill any procurement proposal from the armed forces," and could furthermore "set the hardware-modernization agenda through its power to veto or delay acquisition from overseas in favor of indigenous research and development."[52]

At the same time, the Indian military must bear some of the blame for delays and failures in indigenous weapons programs. It often tries to add new requirements and new capabilities to weapons projects that are already well into R&D, slowing development and deployment and sometimes even leading to the cancellation or scaling back of the program. This then forces the military (or gives it the excuse) to acquire a (often superior) foreign system.[53]

The challenges to the Indian defense industry will likely only increase over the next several years, especially as the country embarks on a massive recapitalization of its armed forces. Estimates are that the military will, over the next two decades, need to buy up to 450 combat aircraft, 100 transport aircraft, 200 helicopters, 1,500 tanks, 500 combat vehicles, 1,500 artillery pieces, and 140 naval ships, including up to 20 submarines and 2 to 3 aircraft carriers.[54] It is arguable whether the Indian military-industrial complex is up to the task of supplying state-of-the-art systems to the nation's armed forces within this timeframe.

Reforming the Indian defense industry, 2001 to the present

To be fair, the Indian government has long been aware of the deficiencies affecting the country's defense industrial base, and for roughly a decade it has pursued a number of initiatives intended to reform and revitalize the defense sector. These reforms generally fall under one of several categories: (1) opening up defense contracting to the private sector; (2) permitting foreign firms to invest in India's defense industry; (3) encouraging more joint R&D and coproduction arrangements with foreign firms; (4) formalizing offsets and leveraging them for technology acquisition; and (5) encouraging arms exports.

In order to shake the state-owned defense sector out its complacency, the Indian government has increasingly invited the commercial sector to compete in defense bidding and production. In 2001 New Delhi allowed private sector participation in defense contracting up to 100 percent of the value of the program.[55] As a result, local commercial firms have begun to win military contracts. Two local firms, Larsen and Toubro (L&T) and Tata, were recently awarded a joint contract to develop components for a new multiple rocket launcher. L&T was also selected to build hulls for India's new nuclear-powered *Arihant*-class submarine (formerly the Advanced Technology Vessel, or ATV), while Tata will produce control system for this sub.[56] In addition, L&T is investing heavily in modernizing its shipyards in Hazira, on India's east coast, in an effort to win away from Mazagon Dock a potentially lucrative follow-on contract to build up to six *Scorpène*-class submarines for the Indian Navy.[57]

For its part, Tata is seeking a tie-up with AgustaWestland of Italy to produce helicopters. Private companies may also bid to build a new armored fighting vehicle for the Indian Army.[58] Altogether, the country's private sector did about US$800 million worth of defense work in 2010, compared to US$4.5 billion earned by the DPSUs and OFs.[59]

In addition, the Indian military hopes to leverage the capabilities of local industry when it comes to commercial off-the-shelf (COTS) solutions, especially when it comes to information technologies (IT). The expectation is that COTS-based solutions would be quicker, more cost-effective, and more easily upgradeable, particularly in areas such as communications, command and control system, situational awareness, and network management.[60]

In order to promote increased commercial participation in defense bidding, the government proposed an initiative in 2007 to designate several private sector companies as "Champions of Industry" (RURs), entitled to the same benefits as DPSUs. RURs would be able to design, develop, and manufacture military equipment, as well as produce defense systems developed by the DRDO. Additionally, they would be eligible for duty-free import of defense research-related equipment, could enter into technology-transfer and licensed-production arrangements with foreign firms, and would receive government military R&D funding. Roughly a dozen local firms applied for RUR status.[61] To further encourage private-sector participation in armaments production, the Indian defense ministry around this same time proposed setting aside 1 billion rupees (US$2.2 million) to fund military R&D projects by commercial firms.[62]

At the same time that the Indian commercial sector was permitted to compete for military contracts, the government also allowed foreign firms to invest in local defense enterprises, up to 26 percent of value. The United Kingdom's BAE Systems, for example, has linked up with Mahindra & Mahindra Ltd., a private Indian conglomerate (one of its divisions builds automobiles and utility off-road vehicles), to develop land defense systems in India. The European Aeronautic Defense and Space Company (EADS, now Airbus) has proposed a defense joint venture with L&T, while Elta of Israel has invested 2.5 billion rupees (US$56 million) in L&T and Astra Microwave to develop and build radar and other defense electronics systems.[63]

Joint ventures between foreign firms and the state-run defense industrial sector have also greatly expanded in recent years. The *Brahmos* cruise missile, for example, is the product of a joint venture between the DRDO and the Russian Federation's NPO Mashinostroyenia. In addition, in 2006, Russia's Irkut Corporation entered into a U$700 million joint venture with HAL to design and build a 60-ton multirole transport aircraft (MRTA).[64] Most important of all, perhaps, Moscow and New Delhi have agreed to codevelop a fifth-generation fighter (FGFA) based on the Russian PAK FA program, which in turn is based on the Sukhoi T-50 prototype. Under the terms of this joint venture, HAL will work with Sukhoi to develop a two-seater version of the T-50, in exchange for a 25 percent work share in the aircraft's design and development, including the mission computer, navigation systems, cockpit displays, and countermeasure systems. The project will also entail considerable Russian technology transfers to India. Altogether, New Delhi could invest up to US$35 billion into the FGFA, including R&D and the procurement of 250 aircraft. Russia and India would also set up a joint marketing company to export this fighter.[65]

Other countries besides Russia are entering into joint ventures with Indian arms producers. Israel Aerospace Industries (IAI) is cooperating with the DRDO to develop a longer-range version of the Israeli *Barak* air-defense missile. Boeing and Tata have set up a joint venture to manufacture defense systems. The French jet engine manufacturer SNECMA is collaborating with HAL on improving the *Kaveri* turbofan engine, and the European missile consortium MBDA is working with the DRDO and BDL to develop a new short-range surface-to-air missile.[66] The United Kingdom's Cobham plc is cooperating with HAL on air-to-air refueling probes for the IAF's Su-30MKI fighters. Finally, regarding the MMRCA program, Dassault was obliged to provide billions of dollars worth of technology transfers in order to enable India to produce the *Rafale* aircraft in India.

In an effort to formalize technology transfer obligations, the Indian government has over the past decade inaugurated and refined an official defense offsets policy.[67] In the 2000s the New Delhi's Defense Procurement Procedures (DPP) guidelines outlined three broad acquisition strategies for the Indian armed forces: "Buy," "Buy and Make," and "Make." "Make" refers to military products that would be more or less wholly designed, developed, and manufactured within India; its basic objective is to ensure the maintenance and expansion of indigenous R&D, design, and production capabilities on the part of the local defense sector, both state-owned and private.[68] The "Buy" category entails products that are intended to be imported; under the terms of the 2006 DPP, any such arms import greater than 3 billion rupees (approximately US$67 million) required a minimum 30 percent direct offset, either in the form of counter-purchases of Indian defense equipment or a foreign direct investment (FDI) in the Indian defense industry (such as codevelopment or coproduction arrangements, or joint international marketing efforts). The "Buy and Make" category applies mainly to major military programs, such as the MMRCA, that entail licensed-production inside India and which therefore demand considerable technology transfers and industrial participation. In such cases, a 50 percent offset is usually mandated. To put it another way, the MMRCA program, which was estimated to be worth at least US$10 billion, was supposed to generate a minimum of US$5 billion in offsets.

India's offset policy reportedly generated around 75 billion rupees (US$1.68 billion) worth of offset work for the period of 2006–2009.[69] Ultimately, the Indian government expects to sign US$10 billion worth of offset agreements over the course of the 11th Defense Plan (2007–2012).[70]

Finally, India has a nascent effort underway to break into the global arms export business. Indian defense industries are increasingly present at major trade shows (such as Farnborough or the Singapore Air Show) in an effort to sell their wares overseas. In particular, India has pitched the *Brahmos* missile, which is marketed by a Russo-Indian joint venture. So far, however, Indian arms exports have been minimal, around US$100 million annually.[71]

In general, therefore, the Indian government is seeking to use private firms to put pressure on the state-owned defense sector to reform itself. By permitting commercial businesses to bid on defense contracts and to create joint ventures with foreign defense companies, it is hoped that the competition will force the DPSUs and OFs to become more market oriented and cost effective, and also more responsive to customer requirements (i.e., the Indian military). In addition, a formalized offsets strategy is intended to inject critical technologies into the Indian military-industrial complex where they are most needed, and in a timely fashion.[72]

Perhaps it is too soon to tell if these initiatives will have their desired impact, but so far, however, these efforts have shown few tangible results. First of all, it has been difficult, for example, to encourage India's private sector to invest in a line of work that requires large, risky investments in R&D and infrastructure in exchange for low returns. This effort is made all the harder by the persistence of a "statist mindset" on the part of the Indian Ministry of Defense – and especially the DRDO – that still tends to default to giving large military contracts (and defense offsets) to the DPSUs and OFs. This was most recently reflected in the defense ministry's 2010 awarding of a noncompetitive contract worth US$1 billion to Bharat Electronics for the Indian Army's battle management systems (BMS).[73] Moreover, the RUR initiative was eventually abandoned, in part due to stiff resistance from trade unions representing workers in the DPSUs and OFs.[74]

Additionally, while the government has permitted foreign firms to invest in Indian defense companies (up to 26 percent of shares), so far there have been few takers. Overseas investors have no independent means by which to valuate these companies' stock, and they are not permitted much say in how these companies should be run. Additionally, the Indian government has frequently rejected foreign shareholding or joint venture efforts, and consequently only four such foreign-Indian defense joint ventures have been set up since 2001.[75] According to one source, in 2009 such FDI amounted to less than US$142,000.[76] At the same time, any privatization of the country's state-owned defense sector, i.e., the DPSUs or OFs, is so unlikely as to be almost inconceivable.

Second, it is still uncertain how much of an impact these new offsets and technology transfers policies may have when it comes to injecting much-needed cutting-edge technologies into the Indian military-industrial complex. For example, in 2015 the whole MMRCA deal fell apart due to cost overruns and squabbles between HAL and Dassault over licensed assembly guarantees. Instead, India abandoned the idea of licensed-production and placed an order for an off-the-shelf buy of 36 Rafale fighters, worth nearly US$9 billion.[77] India's arms producers could be hard-pressed to exploit the foreign technologies they are acquiring if they are unable to also upgrade their capacities for technology absorption, innovation, and production. This could in particular undercut their efforts to make substantive contributions to joint-venture programs such as the FGFA.[78]

Finally, rapidly increasing military expenditures have actually been counter-productive to reforming the state-run defense sector. India's defense budget has grown 60 percent in just the past decade, and analysts expect New Delhi to spend at least US$200 billion on new weaponry over the next 15 years. This is enough to buy nearly every item on the military's "wish-list," as well as to provide a huge windfall of orders that should keep the DPSUs and OFs operating at full capacity for several decades. In light of such an expectation of "fat years" out to the horizon, it will be doubly difficult to encourage state-owned industries to think about being more efficient and market oriented, or to get the Indian military to exert such pressure on the defense sector.[79]

Conclusion

It is, of course, rather easy to be dismissive of Indian's defense industry – both of its current capacities for advanced armaments production and of its likelihood of engaging in real, effective reform. In general, India's track record in both of these areas has not been encouraging, and overall the history of India's military-industrial complex has been particularly disheartening. After China, India possesses the largest and most ambitious defense industrial base in the Asia-Pacific, if not the entire developing world, and yet the performance of its defense industry over the past 50 years has been disappointing in the very least. Billions of dollars have been spent on domestic weapons programs that have never performed up to their requirements or met their objectives when it came to costs and milestones. And while the rest of the world has marveled at India's globally competitive information technologies sector, the country's defense industry has remained, for the most part, an overwhelmingly statist enterprise undauntedly committed to autarkic armaments production. The Indian military-industrial complex was a huge white elephant of highly protected, monopolistic state-owned corporations, headed up by a bloated DRDO, which pressed for indigenous solutions with little heed paid to capabilities or timeliness. It is little wonder, therefore, that Cohen and Dasgupta would flatly state that the Indian military-industrial complex "has not delivered a single major weapon system to the armed forces in five decades of existence."[80] And while such an assertion may strike one as unfair and exaggerated, the sad fact is that this claim was right more times than it was wrong.

All that said, change may be in the offing. Undoubtedly, restructuring and reforming the Indian defense industry will be slow and incremental. At the same time, recent reform efforts have already produced some tangible results. India's private sector has broken finally into the once-restricted arms-producing business, and by 2010 local commercial firms were earning about US$800 million annually from defense contracting.[81] Moreover, private-sector bidding for local defense contracts is likely only to grow as these companies increase their investments in capabilities and facilities for armaments production, such as shipbuilding, military vehicles, and defense-related electronics. In mid-2015, for example, Tata and Airbus were permitted to establish a joint venture to produce C-295 transport aircraft in India.[82] In addition, opening up the military contracting process to foreign firms through joint ventures and offset arrangements is also fundamentally altering the defense-industrial landscape of India.

Defense industrial reforms also have some powerful allies in the government and the military. In particular, both are keen to use the local private sector and foreign firms' involvement to pressure the DRDO, DPSUs, and OFs to change their business-as-usual practices. In this regard they are strongly supported by powerful allies such as the Confederation of Indian Industries (CII), which has long pressed for the liberalization and opening up of the country's defense business. In addition, the Modi government has been particularly supportive of defense reform: in its first year in power it has fast-tracked 40 acquisition projects worth over a billion rupees (US$14 billion), the bulk of which come under "Buy" or "Buy and Make" categories; at the same time, Modi is "tightening the screws" on the DRDO and state-owned defense firms "to ensure they deliver on time in a cost-effective manner."[83]

Nevertheless, reforming India's military-industrial complex remains an uphill battle. The state-owned defense sector is still very powerful, and the DPSUs and OFs will likely continue to strongly oppose any initiatives to remove or reduce their role as the primary producers of the nation's armaments, particular when it comes to such big-ticket items as combat aircraft, warships, missile systems, tanks, and other armored vehicles. Moreover, the DRDO still wields considerable influence within the national armaments planning process, and is thus a strong advocate for the *status quo*. In particular, it still prefers, when it can, to pursue indigenous development programs over licensed-production or foreign joint ventures.

Ultimately, it is too soon to tell if these recent reform efforts will take root and flower. One thing is for certain, however: so long as India continues to shield and coddle its traditional military-industrial complex, to pursue a Nehruvian socialist style of ownership and operation, and to emphasize self-sufficiency as a strategic political-economic goal over military requirements and capabilities, the local defense industry will continue to be challenged when it comes to supplying the Indian armed forces with the equipment it requires.

Notes

1 Recent studies on the Indian defense industry include Stephen P. Cohen and Sunil Dasgupta, *Arming without Aiming: India's Military Modernization* (Washington, DC: Brookings Institution, 2010); Deba R. Mohanty, *Arming the Indian Arsenal* (New Delhi: Rupa, 2009); Ajay Singh, "Quest for Self-Reliance," in Jasit Singh, ed., *India's Defense Spending* (New Delhi: Knowledge World, 2000); Deba R. Mohanty, *Changing Times? India's Defense Industry in the 21st Century* (Bonn: Bonn International Center for Conversion, 2004); Rahul Bedi, "Two-Way Stretch," *Jane's Defense Weekly*, February 2, 2005; Manjeet S. Pardesi and Ron Matthews, "India's Tortuous Road to Defense-Industrial Self-Reliance," *Defense & Security Analysis*, Vol. 23, No. 4 (December 2007); Timothy D. Hoyt, *Military Industry and Regional Defense Policy: India, Iraq, and Israel* (New York: Routledge, 2007).
2 Singh, "Quest for Self-Reliance," pp. 126–127.
3 Singh, "Quest for Self-Reliance," p. 127.
4 Angathevar Baskaran, "The Role of Offsets in Indian Defense Procurement Policy," in J. Brauer and J.P. Dunne, eds., *Arms Trade and Economic Development: Theory, Policy, and Cases in Arms Trade Offsets* (London: Routledge, 2004), pp. 211–213, 221–226.
5 Pardesi and Matthews, "India's Tortuous Road to Defense-Industrial Self-Reliance," pp. 421–429. See also the chapter in this same volume by Ron Matthews and Alma Lozano, "India's Defense Acquisition and Offsets Policy."
6 Richard A. Bitzinger, *Towards a Brave New Arms Industry?* Adelphi Paper No. 356, International Institute for Strategic Studies (Oxford: Oxford University Press, 2003), pp. 16–18.
7 Singh, "Quest for Self-Reliance," p. 151.
8 Laxman Kumar Behera, "India's Growing Defense Industry Base," *Defense Review Asia*, November 2010, p. 34.
9 "Interview with M. Mangapati Pallam Raju, Minister of State for Defense, India," *Defense News*, March 7, 2011; Behera, "India's Growing Defense Industry Base," pp. 31–34.
10 Laxman Kumar Behera, "Background Paper on India's Defense Industry: An Overview" (prepared for the "National Seminar on the Defense Industry," New Delhi, January 23–24, 2009), p. 4.
11 Behera, *Background Paper on India's Defense Industry*, p. 4.
12 Vivek Raghuvanshi, "Exec: Award Shows India Still Favors State Firms," *Defense News*, October 4, 2010.
13 Behera, "India's Growing Defense Industry Base," p. 34.
14 Brian Cloughly, "Analysis: DRDO Fails to Fix India's Procurement Woes," *Jane's Defense Weekly*, June 28, 2010.
15 Vivek Raghuvanshi, "Indian Research Agency Agrees to Tech Transfers," *Defense News*, January 25, 2010.
16 ADA website (http://www.ada.gov.in).
17 Author's interviews in India, March 2011.
18 Stockholm International Peace Research Institute (SIPRI), *SIPRI Military Expenditure Database*, 2011 (https://www.sipri.org/databases/milex).
19 Vivek Raghuvanshi, "Budget Hike in India," *Defense News*, March 7, 2011.
20 Vivek Raghuvanshi, "Report: Indian Products Defective," *Defense News*, January 9, 2006.
21 Author's interviews in India, March 2011; see also Mohanty, *Changing Times?*, pp. 28, 36–37; Pardesi and Matthews, "India's Tortuous Road to Defense-Industrial Self-Reliance," pp. 432–434; Baskaran, "The Role of Offsets in Indian Defense Procurement Policy," pp. 213, 216–218.
22 Cloughly, "Analysis: DRDO Fails to Fix India's Procurement Woes."

23 Shiv Aroor and Amitav Ranjan, "23 Yrs and First Fighter Aircraft Hasn't Taken Off," *Indian Express*, November 14, 2006.

24 Shiv Aroor and Amitav Ranjan, "Arjun, Main Battle Tanked," *Indian Express*, November 14, 2006; Laxman Kumar Behera, "The Saga of MBT-Arjun," *Defense Review Asia*, June 2010, pp. 20–22.

25 Government of India, press release, "Arjun Battle Tank," May 5, 2008 (http://pib.nic.in/newsite/erelease.aspx?relid=38445).

26 Cohen and Dasgupta, *Arming without Aiming*, p. 33.

27 Joshi, "If Wishes Were Horses"; Shiv Aroor and Amitav Ranjan, "Armed Forces Wait as Showpiece Missiles are Unguided, Way Off Mark," *Indian Express*, November 13, 2006.

28 Ajal Shukla, "Army Opts for Nag Missile as it Enters Final Trials," *Business Standard*, March 8, 2010; Gordon Arthur, "Indian Armed Force Programs: Large Budget Increases," *Defense Review Asia*, March 2009, pp. 13–14.

29 Bedi, "Two-Way Stretch."

30 Raghuvanshi, "Report: Indian Products Defective."

31 Six aircraft originally competed for the MMRCA: the US F-16 and F/A-18, the Russian MiG-35, the Swedish *Gripen*, the French *Rafale*, and the Eurofighter *Typhoon*. In May 2011 the IAF shortlisted the *Rafale* and *Typhoon*, and a finalist will be announced probably in 2012.

32 Arthur, "Indian Armed Force Programs," pp. 13–14.

33 Arthur, "Indian Armed Force Programs," p. 14.

34 Aroor and Ranjan, "Arjun, Main Battle Tanked"; Manu Pubby, "Israel, EU in Contention to Co-develop Radars for *Tejas*," *Indian Express*, July 14, 2010.

35 Author's interviews in India, March 2011; Globalsecurity.org, "PJ- 10 BrahMos" (http://www.globalsecurity.org/military/world/india/brahmos.htm).

36 The IAF will acquire the first 18 MMRCA aircraft directly from the foreign manufacturer; all subsequent aircraft will be licensed-produced in India.

37 Author's interviews in India, March 2011.

38 Singh, "Quest for Self-Reliance," p. 151; Bedi, "Two-Way Stretch."

39 Cloughly, "Analysis: DRDO Fails to Fix India's Procurement Woes"; Guy Anderson, "India's Defense Industry," *RUSI Defense Systems*, February 2010, p. 68.

40 Stockholm International Peace Research Institute (SIPRI), *SIPRI Arms Transfers Database* (http://www.sipri.org/databases/armstransfers/armstransfers); Rahul Bedi, "India Announces 12% Defense Budget Increase," *Jane's Defense Weekly*, March 3, 2011.

41 Bedi, "Two-Way Stretch," pp. 25–26; Singh, "Quest for Self-Reliance," p. 155.

42 Bedi, "India Announces 12% Defense Budget Increase."

43 Pardesi and Matthews, "India's Tortuous Road to Defense-Industrial Self-Reliance," p. 424.

44 Pardesi and Matthews, "India's Tortuous Road to Defense-Industrial Self-Reliance," pp. 432–434; Singh, "Quest for Self-Reliance," pp. 148–149.

45 Author's interviews in India, March 2011.

46 Quoted in Bedi, "Two-Way Stretch," p. 28.

47 Bedi, "Two-Way Stretch," p. 26.

48 Bedi, "Two-Way Stretch," p. 27; Rahul Bedi, "India Launches 'Thorough' Audit of DRDO's Effectiveness," *Jane's Defense Weekly*, January 24, 2007; Rahul Bedi, "Making Decisions," *Jane's Defense Weekly*, January 25, 2010; Manoj Joshi, "If Wishes Were Horses," Hindustan Times, October 18, 2006.

49 Cohen and Dasgupta, *Arming without Aiming*, p. 33.

50 Joshi, "If Wishes Were Horses."

51 Cohen and Dasgupta, *Arming without Aiming*, p. 33.

52 Cohen and Dasgupta, *Arming without Aiming*, p. 33.

53 The author is grateful to Timothy Hoyt for pointing this out.

54 Vivek Raghuvanshi, "India May Increase Defense Spending as Percent of GDP," *Defense News*, September 24, 2007; Bedi, "Two-Way Stretch," p. 28; author's interviews in India, March 2011.

55 Rahu Bedi, "Eyeing the Prize," *Jane's Defense Weekly*, January 17, 2007.

56 Josy Joseph, "Private Sector Played a Major Role in Arihant," *Daily News & Analysis*, July 27, 2008 (http://www.dnaindia.com/india/report_private-sector-played-a-major-role-in-arihant_1277435).

57 Jon Grevatt, "L&T Chief Attacks Indian Government Support for 'Sick' Public-Sector Companies," *Jane's Defense Industry*, March 9, 2011.

58 Vivek Raghuvanshi, "Private Firms to Bid for Indian Vehicle Project," *Defense News*, August 23, 2010.

59 Vivek Raghuvanshi, "Tata Seeks 'Level Playing Field' in India," *Defense News*, February 2, 2011.

60 Author's interviews in India, March 2011.

61 Anderson, "India's Defense Industry," p. 69; Bedi, "Eyeing the Prize."

62 Rahul Bedi, "India Plans Private Sector Defense R&D Project Funding," *Jane's Defense Weekly*, August 1, 2007.

63 Anderson, "India's Defense Industry," pp. 69–70.

64 Bedi, "Eyeing the Prize."

65 Author's interviews in India, March 2011; Ajal Shuka, "India, Russia close to PACT on next generation fighter," *Business Standard*, January 5, 2010 (http://www.business-standard.com/india/news/india-russia-close-to-pactnext-generation-fighter/381718); Ajal Shuka, India to develop 25% of fifth generation fighter," *Business Standard*, January 6, 2010 (http://www.business-standard.com/india/news/india-to-develop-25fifth-generation-fighter/381786).

66 Author's interviews in India, March 2011; Pierre Tran, "MBDA Looks to India for New Air Defense Missile," *Defense News*, February 12, 2009.

67 See Matthews and Lozano, "India's Defense Acquisition and Offsets Policy," for a fuller explanation of India's new defense offsets policies.

68 Behera, "India's Growing Defense Industry Base," p. 36.

69 Anderson, "India's Defense Industry," p. 69.

70 Behera, "India's Growing Defense Industry Base," p. 36.

71 Anderson, "India's Defense Industry," p. 68.

72 Author's interviews in India, March 2011.

73 Raghuvanshi, "Exec: Award Shows India Still Favors State Firms."

74 Jon Grevatt, "India Delays Defense Reforms Again in Face of Multiple Pressures," *Jane's Defense Weekly*, December 21, 2007; Anderson, "India's Defense Industry," p. 69.

75 Author's interviews in India, March 2011.

76 Anderson, "India's Defense Industry," p. 69.

77 Vivek Raghuvanshi, "India and France to Finalize $8.9 Billion Deal for 36 Rafales," *Defense News*, April 19, 2016.

78 Bedi, "Making Decisions."

79 Author's interviews in India, March 2011.

80 Cohen and Dasgupta, *Arming without Aiming*, p. 32.

81 Vivek Raghuvanshi, "Tata Seeks 'Level Playing Field' in India," *Defense News*, February 2, 2011.

82 Vivek Raghuvanshi, "Airbus, Tata to build Indian AF transport," *Defense News*, May 15, 2015.

83 Rajat Pandit, "In first year, Modi govt cleared defense proposals worth Rs 1L crore," *Times of India*, May 16, 2015.

5   The Republic of Korea

The Republic of Korea (ROK or South Korea) has aggressively pursued a "domestic weapons first" policy going back to the early 1970s and the implementation of the first Yulgok Project, an ambitious program of defense industrialization that was intended to lay down "a basic foundation for a self-defense capability for the 21st century."[1] This indigenization process was initially propelled by the threat from North Korea, and the belief that achieving self-sufficiency in defense procurement was essential to maintaining an adequate defense capability. At the same time, domestic arms production was more than merely achieving "security of supply;" very powerful technonationalist impulses can be detected in South Korea's defense-industrialization activities over the past several decades. In case of the ROK, defense-industrial technonationalism had three core objectives: first, to strengthen its national political independence by reducing dependency upon foreign sources of arms; second, to aid domestic economic development overall by pursuing armaments production as an import-substitution strategy and as a driver of technology-intensive industrialization; and third (and perhaps most important all), to enhance the nation's military-political status and raise its profile as an important geopolitical player in Asia.

Such technonationalism, when it comes to South Korean armaments production, can be traced back to the regime of Park Chung Hee (1961–1979) and his policy of "Self-Reliant National Defense." Reducing the country's dependencies on foreign armaments is a critical national security objective, but for reasons that go far beyond national defense. In the first place, South Korea is committed to a strategy of "cooperative self-reliant defense,"[2] an approach that embraces the US-ROK alliance (as embodied in the 1953 Mutual Defense Treaty), but which also seeks to strengthen the country's national defense capabilities –[3] that is, "acquiring the ability to independently develop primary weapon systems for core force capability."[4] This serves as least two military functions. First, having an indigenous defense industry permitted South Korea to tailor weapons systems especially suited for its unique national defense requirements and to take advantage of technological breakthroughs in its own domestic R&D base. Second, local armaments manufacturing can be used as a bargaining chip in joint ventures with foreign arms producers, that is, a means by which to extract industrial or technological concessions, such as technology transfers or production offsets, when it comes to arms imports; this further advances South Korea's efforts to wean itself off its defense-industrial dependencies on larger powers, particularly the United States.

At the same time, defense industrialization was also viewed as at least partially driving the expansion and technological modernization of the overall national economy. Armaments production was a means by which to stimulate the development of new industrial sectors and to underwrite research and development when it came to new technologies, such as in the area of aerospace and electronics. Advanced military technologies were particularly seen as contributing to the growth of import-substituting industries. These military-industrial developments, in turn, could realize important commercial gains. In some cases these "spin-off" benefits were direct – such as heavy trucks or the MD-500 helicopter, which were produced in both military and civilian versions; in other cases they could be indirect, such as using military aerospace production as the basis for establishing and nurturing a commercial aircraft industry.[5] Either way, military industrialization went hand in hand in South Korea's overall development strategies.

Besides all these aforementioned arguments, however, South Korea's striving for self-reliance in armaments has been as much about expanding the country's autonomy and freedom of action in foreign affairs and regional politics. The "rich nation/strong army" argument provides a useful insight into South Korea's decision to invest heavily in the domestic defense industry. Possessing an independent defense industrial capability feeds directly into their concepts of national power – not only by creating military power but also by demonstrating the country's industrial and technological prowess, and thereby confirming its status as a great power in the broadest sense.

In the first place, possessing a strong and technologically advanced domestic defense industry was seen as a demonstration of the country's emergence as a high-technology powerhouse and therefore as a rising technological-industrial-economic power overall to be reckoned with. Second, the ROK's drive to expand its indigenous arms production capabilities is congruent with its rising geopolitical ambitions in the region; Seoul's "future-oriented" aspirations for regional great-power status closely correlates to Samuel's "rich nation/strong army" technonationalism. On the one hand, South Korea seeks to position itself as a "full-fledged player upon the regional stage," especially after reunification, when a "united Korea faces the superior forces of Japan, China, and Russia."[6] At the same time, Seoul perceives its long-term future in Asia as a "regional balancer,"[7] a "middle power amid the growing rivalry between China and the US in the Asia-Pacific."[8] This, in Seoul's view, requires a certain degree of autonomy from other powers, particularly when it comes to national security.[9] Consequently, "the Koreans desire to acquire air and naval weapons that can compete with the high-technology weapons of these larger and more developed states."[10]

Third, autarky in armaments is seen to be a critically needed hedge against being overly dependent on the United States, "a patron-state it either sees as potentially unreliable in the long term, or with whom it is increasingly at-odds over fundamental military and geopolitical issues (such as the salience of the North Korean threat)."[11] As Bitzinger and Kim point out:

> For many reasons, South Korea has never been completely at ease over its dependence on the United States as its major arms supplier. In the first place, these dependencies left Seoul vulnerable to unilateral changes in U.S. policy toward the Korean peninsula. In addition, Washington could – and indeed some- times did – hold back the release of weapons systems and technologies that Seoul viewed as critical to its self-defense. Third, the United States' position as a near-monopolistic supplier to South Korea led to suspicions that it was overcharging Seoul for military equipment. Finally, many South Koreans regard Washington's arguments about deploying joint weapon systems in the interests of interoperability to be simply a "thin-veiled blackmail tactic."[12]

Bitzinger and Kim add that, "given the Korean peoples' deeply embedded distrust of big powers, the development of a 'self-defense posture' – and the corollary concept of possessing a strong domestic arms industry – has become a national obsession for South Korea."[13] Consequently, South Korea's fixation with building up its indigenous defense industry is a perfect example of Samuel's technonationalistic, "rich nation/strong army" paradigm. In particular, in Seoul's mind securing a higher degree of self-sufficiency in armaments is a critical necessity if South Korea, the self-perceived historical underdog, is to move from geopolitical victim to regional player.

Development and growth of the South Korean defense industry

South Korea only began to produce arms in significant quantities and types in the early 1970s; before then the country had relied mainly upon the United States for its military equipment. ROK military manufacturing initially centered around the production of small arms and ordnance, such as automatic rifles, grenades, mortars, and mines, along with uniforms, tents, mess-kits, and the like. Local arms production quickly expanded, however, and by the 1990s South Korea had already built up a considerable military manufacturing capacity. By the second decade of the 21st century the ROK possesses one of the most sophisticated arms industries among the newly industrialized countries, capable of manufacturing a wide array of indigenous weapon systems, including combat aircraft, tanks and other armored vehicles, major surface combatants, submarines, and even some types of missile systems. As a result, South Korea has attained a high degree of self-sufficiency in defense acquisition.

The rapid expansion of the ROK's arms industry was aided by three factors: (1) extensive foreign assistance, (2) the presence of an already well-established, domestic heavy industrial base, and (3) the strong and instrumental role that the central government played in launching and nurturing the process of domestic armaments production. Initially at least, local armaments production relied heavily upon the licensed-production of foreign military systems, or upon components, technologies, and designs imported from abroad, particularly the United States. Foreign technology inputs and assistance enabled the South Korean defense industry to achieve a high degree of self-sufficiency in production in a short period of time and with relatively low entry costs.[14] In this regard the implementation of the Nixon Doctrine was a real boost to the South Korean defense industry, for while it reduced US defense commitments to Asia, at the same time it also liberalized the export of advanced military technologies to allies such as the ROK.

In addition, the development of South Korea's arms industry proceeded in conjunction with the country's overall economic modernization and industrialization goals. Like other developing countries, the ROK consciously pursued a parallel strategy of "security and development," that is, of building up its heavy industry and high-technology sectors at the same time as it strove for self-sufficiency in arms production. Particularly during the Park regime, South Korea's leadership viewed economic development to be an essential element of national security, and state investments in such heavy industry sectors as steel, machinery, transportation, and chemicals were regarded as directly contributing to defense preparedness.[15] Industrialization and technological advancement were seen as feeding into the development of domestic arms-manufacturing capabilities, such as building up general skills and know-how, and in providing lead-in support or equipment for arms production. The establishment of the South Korean commercial shipbuilding industry, for example, facilitated the construction of warships, while the creation of a domestic iron and steel industry and later a domestic automobile industry provided the parts and materiel (such as armor plating, chassis, and engines), as well as the skilled labor necessary for the production of military vehicles.

Finally, the South Korean government has been heavily involved in the arms production process by providing direct and indirect subsidies to manufacturers, by underwriting defense research and development planning, and by designating firms as sole-source suppliers of critical military equipment.[16] In the first place the central government has underwritten the Korean defense industry by virtue of it (that is, the ROK military) being the prime consumer of locally produced arms. It guarantees procurement and therefore profitability; it has also created "monopolistic or oligopolistic positions with the defense sector, as each individual contractor has been directed to specialize in a certain area of production."[17] In addition, Seoul has encouraged firms to enter into domestic arms production through a variety of incentives – such as tax breaks, low-interest loans, and direct financial support – as well as coercive measures, such as tying defense contracting to state support for engaging in other types of commercial production.[18]

The ROK government has traditionally assumed most of the risk when it comes to weapons development. While most armaments production in Korean is undertaken by the private sector (see further on), most defense R&D is, for the most part, managed by the government-run Agency for Defense Development (ADD). The ADD operates at the center of the national defense R&D process in South Korea. It has primary responsibility for the research and development of indigenous weapons systems and core technologies, manages the development of dual-use and core technologies, and carries out operational testing and evaluation of developmental systems. It is directly responsible to the Ministry of National Defense's Defense Acquisition Program Administration (DAPA), which oversees armaments acquisition in South Korea, including determining requirements, approving R&D projects, and assessing testing and evaluation results. Particular emphasis is put on supporting local defense industries and armaments programs, promoting the export of Korean weaponry, and protecting homegrown defense technologies. The ADD works directly with the local defense industry on proto-typing and producing ADD-developed weapons systems, as well as with industry think tanks, universities, and research institutes on basic and applied research and on core technology development.

The ADD has a staff of approximately 2,500, of which 84 percent are engi-neers, technicians, scientists, and other personnel engaged in research and develop-ment.[19] The ADD comprises seven R&D institutes (precision-guided munitions, command, control, communications and computing [C⁴], intelligence, surveillance and reconnaissance [ISR], so-called "neo-technologies," ground systems, naval sys-tems, and aircraft systems) and one test center. Each ADD R&D institute operates its own network of research laboratories. South Korea typically allocates approxi-mately 5 percent of its defense budget to military R&D, which would translate into a defense R&D budget of approximately US$1.67 billion in 2015.[20]

South Korean arms manufacturing began modestly enough. In 1971 the ROK was permitted to license-produce the American-designed M16 rifle. This was subsequently followed by licenses to assemble or coproduce other weapon sys-tems, mostly from the United States. During the 1970s and 1980s this included the F-5 fighter jet, the MD-500 helicopter, M101 and M114 howitzers, and the PSMM-5 fast attack boat. This was followed in the 1980s by the licensed-production of the UH-60 Blackhawk helicopter and the M-109 155mm self-propelled howitzer.

Starting in the late 1980s and early 1990s, the ROK began to undertake more sophisticated types of licensed-production. Seoul negotiated sizable "offsets" – i.e., coproduction rights – for several of its foreign arms acquisitions. For exam-ple, when South Korea decided to purchase 40 F-16 fighters from the United States in the mid-1980s, it demanded and received the right to produce several of the plane's key subsystems, including center fuselage section, cockpit side panels, and ventral fins.[21] This program was followed up by an agreement for an addi-tional 140 F-16s that included the establishment of a complete turnkey manu-facturing facility in South Korea in order to build the jet locally. Under the terms of this arrangement, designated the Korean Fighter Program (KFP), the South Koreans purchased 12 F-16s directly from the United States, then assembled 36 F-16s from U.S.-supplied kits, and eventually produced 92 additional fighters using both Korean- and U.S.-built components.

Also during the 1990s the ROK has begun to diversify its sources of foreign technology away from an overwhelming dependency upon the United States. Similar coproduction arrangements were made for Hawk trainer jets acquired from the United Kingdom, for Type-6614 wheeled armored personnel carriers from Italy, and for Bv-206 all-terrain tracked carriers from Sweden. During the same period Seoul secured an agreement with Germany to construct nine Type-209 submarines, which was subsequently followed up by a similar contract for the manufacture of nine more advanced Type-214 submarines.

More importantly, beginning in the 1990s the process of *indigenous* development and design, as well as the local manufacture of weapon systems, began to take on extra momentum, and gradually the ROK armed forces began to phase out foreign equipment and to replace them with locally produced systems. Beginning in the mid-1980s South Korea had begun manufacturing its own assault rifle, the K2, as well the K200 Korean Infantry Fighting Vehicle (KIFV). By the mid-1990s the ROK was producing the indigenously designed K1/K1A1 main battle tank, which was subsequently replaced by the more advanced K2 Black Panther tank in the early 21st century. Other homegrown weapons systems included the *Chunma* (Pegasus) surface-to-air missile, the K21 infantry fighting vehicle, and the K9 Thunder 155mm self-propelled howitzer.

South Korea's sizable shipbuilding industry has constructed a number of domestically designed warships for the ROK Navy (ROKN), many of them quite innovative in their own right. For over a decade South Korean shipyards have turned out series of progressively more complicated and capable destroyers, known as the KDX (Korean Destroyer Experimental) class. The most recent variant is the 7,700-ton KDX-III, also known as the *King Sejong the Great*-class, of which three have been acquired by the ROKN, and three more are currently under construction. The KDX-III is equipped with the US-supplied *Aegis* air-defense radar and fire-control system and utilizes the Standard SM-2 Block IIIB air-defense missile.[22] As with other *Aegis*-equipped warships, the KDX-III could be upgraded to the Standard SM-3 missile for anti-tactical ballistic missile operations, although this is currently not being planned. Other armaments include the *Hyunmoo*-IIIC land-attack cruise missile (LACM) and either the Harpoon or the indigenous *Haesung* (Sea Star) anti-ship cruise missiles (ASCM), all housed in 128 vertical-launch system (VLS) cells.[23] In addition, the ROKN has taken delivery of at least 2 locally built 14,000-ton *Dokdo*-class amphibious assault ship, each of which is capable of carrying 700 troops, 10 tanks, 15 helicopters, and 2 air-cushioned (hovercraft) landing craft.[24]

In addition, South Korea is building up its domestic submarine industry. It is building three submarines (based on the *Changbogo*-class/Type-209 submarines produced under license from Germany) for export to Indonesia. It is currently constructing 9 German-designed 1,800-ton Type-214 submarines for the ROKN, which feature fuel cell technology for air-independent propulsion (AIP).[25] It is speculated that the ROK will eventually attempt to design and build its own class of submarine, a 3,000-ton boat with AIP and the ability to fire land-attack cruise missiles; a nuclear-powered version has even been mooted.[26]

Perhaps the most noteworthy progress toward autarky that has taken place in South Korea has occurred in the domestic aerospace industry. Originally an assembler of foreign systems, such as the F-5 and F-16 fighter jets and the MD-500 helicopter, the ROK aircraft sector has gradually emerged to become a center of combat aircraft design and development. The ROK aerospace industry's first indigenous product was the KT-1 *Woongbi*, a turboprop basic trainer/light-attack aircraft developed under the auspices of Daewoo Aerospace. Initiated in the late 1980s, the KT-1 had its maiden flight in November 1991, and production started in 1999. Eventually, 105 *Woongbi*s were built for the ROK Air Force (ROKAF). The KT-1 was followed by the T-50 Golden Eagle, an even more ambitious program to design and manufacture a supersonic advanced trainer/light attack jet. The T-50 is Korea's first indigenous jet aircraft, intended to replace T-38, A-37, and F-5 fighters in the ROKAF, as well as export sales. Launched in the mid-1990s, the plane was originally a joint venture between Samsung Aerospace and Lockheed Martin, with the US company supplying critical technologies relating the aircraft's wing, computerized flight-control system, and avionics suite. The T-50 first flew in August 2002, and the aircraft entered service with the ROKAF in 2005. Later versions include the TA-50, a lead-in fighter/trainer/attack plane, and the FA-50, a dedicated fighter aircraft outfitted with more advanced avionics and capable of employing a broader suite of weapons. The ROKAF is likely to acquire up to 140 T-50, TA-50, and FA-50 versions of this aircraft. Other Korean indigenous aerospace products include the KUH-1 *Surion* utility helicopter, a joint project with Airbus Industries' Eurocopter division, and the *Hyunmoo*-III long-range anti-ship/land-attack cruise missile.[27]

Consequently, by the second decade of the 21st century the ROK has become largely self-reliant in most major weapons systems. The local arms industry is particularly broad-based in scope, aided by sizable investments in the aerospace, land ordnance systems, and shipbuilding sectors. Nearly 80 percent of South Korea's arms are procured domestically, including combat aircraft, main battle tanks, armored vehicles, warships, and submarines, and it is becoming increasingly self-reliant in missile systems.[28]

Korea continues to have ambitious plans for its indigenous defense industrial base. Like other countries in the Asia-Pacific region, the modernization of its armed forces is a high priority, and the ROK military has plans to procure several types of advanced weapon systems over the next several decades, including airborne early warning and signals-intelligence aircraft, stand-off munitions, unmanned aerial vehicles, night-vision goggles, and a ballistic missile defense system.[29] In particular, Seoul wants its domestic aerospace industry to design and manufacture a next-generation fighter jet, designated the KFX. Overall, therefore, as Seoul seeks to procure more cutting-edge weaponry, it will place greater demands on its domestic arms industry.

Like Japan, Seoul has traditionally relied upon the private sector rather than state-owned enterprises in order to carry out national arms production – although the central government, through such state-run financial institutions as the Korea Development Bank (KDB) and the Financial Services Commission, holds considerable shares in some arms-producing companies, such as Korea Aerospace Industries (27 percent) and Daewoo Shipbuilding and Marine Engineering (50.4 percent). Nearly 100 local companies are engaged in some kind of defense contracting, supported by 300 or so subcontractors employing an estimated 25,000 workers.[30] In reality, however, nearly all of the ROK's armaments production has long been concentrated in just a handful of large industrial conglomerates, or *chaebol*, particularly Hanwha, Hyundai, LG, Kia, and Daewoo. Hyundai Rotem manufactures the K1 and K2 main battle tanks, for example; Doosan DST (formerly Daewoo Heavy Industries) produces armored vehicles, including the K21 IFV; Kia Motors builds military trucks. LIG Nex1 (a subsidiary of LG Corporation) produces missile systems and electronics, while Hanwha Techwin (formerly Samsung Techwin) manufacturers jet engines and artillery systems (such as the K9 Thunder self-propelled howitzer and the *Chunmoo* multiple rocket launcher).[31] Hyundai Heavy Industries (HHI) and Daewoo Shipbuilding and Marine Engineering (DSME) share construction of KDX-class destroyers and production of the new *Son Wong-il*-class (Type-214) submarine, while HHI and STX Offshore and Shipbuilding are building the *Incheon*-class/FFX frigate for the ROKN.

Two other local defense companies are worth noting: the first is Hanwha Thales, a joint venture between Hanwha Techwin and the French defense firm Thales; Hanwha Thales manufactures a wide array of defense electronics, including command and control systems, surveillance and reconnaissance systems, and optoelectronics, avionics, and radar. The other company is Korea Aerospace Industries (KAI), Korea's leading aircraft manufacturer. KAI was created in 1999 by the forced merger of three money-losing enterprises: Samsung Aerospace (now Hanwha Techwin), Daewoo Heavy Industries Aerospace Division (now Doosan), and Hyundai Space and Aircraft Company. Hyundai and Hanwha[32] each own 10 percent of the company, and Doosan owns 5 percent; the state-owned KDB controls another 27 percent of KAI's shares, with the remainder spread among a variety of other investors. KAI produces all of the country's military aircraft, particularly the T-50 Golden Eagle advanced trainer/lightweight fighter jet, the KT-1 turboprop trainer, and the *Surion* utility-lift helicopter. Other KAI products include unmanned aerial vehicles (UAVs) and components for F-15K fighter aircraft acquired by the ROKAF during the 2000s, and services such as performing depot maintenance for ROKAF aircraft.

In a move that is particularly important for the future of KAI, the company was recently selected to be the lead contractor to build the ROKAF's next indigenous combat aircraft. In March 2015 DAPA awarded a contract to joint KAI-Lockheed Martin team to begin development of the new KFX (Korean Fighter Experimental) future fighter jet. The KFX will have stealthy features similar to the US-built F-35 Joint Strike Fighter (JSF); it will also feature an advanced radar and avionics. The KAI/Lockheed team is pressing ahead with two designs for the KFX: one, a twin-engine, canard-type fighter, and the second a single-engine aircraft closely resembling the JSF. The ROKAF plans to procure as many as 120 KFX fighters, which will replace the air force's aging fleet of F-4 Phantoms and F-5s. Seoul has also succeeded in signing up Indonesia as a partner in the KFX program, and Jakarta could acquire as many as 50 fighters to meet its own "IFX" requirements. Under the terms of the project contract, the KAI-Lockheed team will underwrite 20 percent of incurred development costs, while the ROKAF and Indonesia will cover 60 percent and 20 percent of the costs, respectively. Overall, the KFX could be worth as much as 8.6 trillion won (US$7.8 billion).[33]

In addition to KAI, the Korean Air Lines Co. (KAL) – which manufactured the F-5 fighter and the MD-500 and UH-60s helicopter back in the 1970s and 1980s – continues to run extensive depot maintenance and upgrade facilities, as well as to develop and produce UAVs. Rounding out this list of leading domestic arms manufacturers are Poongsan, which produces ammunition; S&T Dynamics, which manufacturers the K2 assault rifle, machine guns, and 20mm cannon; and Hanjin Heavy Industries, which is constructing the *Dokdo*-class amphibious assault ship for the ROK Navy.

In more recent years the ROK Ministry of National Defense (MND) has encouraged the private sector to become more active in conducting military-related R&D (including investing their own monies), and has directed South Korean industry to specialize in a particular area of military production in the hope of increasing and focusing South Korean military-technical activities at the industry level.[34] In the case of KAI, for example, besides being awarded contracts to build both the KT-1 and T-50 combat aircraft for the ROKAF, the government has pledged that KAI will be granted exclusive rights to future ROK military aircraft contracts, particularly the new KFX fighter jet program.[35] ROK military planners are also seeking to exploit indigenously available advanced technologies through increased spin-on; as such, private firms such as KAI are being encouraged to engage in civil programs like commercial passenger jets. MND hopes that working on these projects will not only make companies like KAI stronger in the global aerospace market, but also develop advanced civilian technologies that could be applied to military production.

South Korea has also emerged as a significant arms exporter in recent years, giving an additional boost to its indigenous defense industry. Initially, the ROK produced military equipment mainly to meet domestic needs, i.e., as an import-substitution strategy and not as an instrument of export-led growth. However, since the turn of the century at least, both the South Korea government and the local arms industry have aggressively pursued arms exports as an invaluable additional source of revenue. Despites its best efforts, however, for a long time the ROK experienced only modest and sporadic success when it came to overseas arms sales. Exports often consisted mostly of low-tech items, such as uniforms and other general military equipment, small arms and ammunition, trucks, and small patrol boats. It scored one or two sizable deals, such as an agreement with Malaysia for approximately 110 K200 infantry fighting vehicles, but there was nothing sustainable in the form of follow-on sales. Indeed, throughout the 1990s South Korean arms exports never exceeded US$75 million per year, according to the Stockholm International Peace Research Institute (SIPRI).[36] Even by the ROK government's own statistics, overseas arms sales remained weak until the middle of the first decade of the 21st century. However, starting around 2007 South Korea began to experience sustained success when it came to arms exports, reaching almost US$240 million in 2011 and 2012.[37] Major sales include (1) KT-1 primary trainer planes to Indonesia, Peru, and Turkey, (2) K9 self-propelled guns to Turkey and Poland, (3) three Improved *Changbogo*-class submarines to Indonesia, (4) *Makassar*-class landing platform dock (LDP) amphibious warfare ships to Indonesia, Peru, and the

Philippines, and (5) one stealthy, multirole frigate to Thailand. In addition, Turkey will coproduce a variant of the K2 Black Panther tank based on a major technology-sharing agreement with South Korea.[38] In particular, South Korea has broken into the highly competitive global fighter market by concluding export agreements with Indonesia, Iraq, and the Philippines for the sale of the T-50 supersonic trainer/light attack jets; altogether, these three countries are buying 52 various versions of the T-50.[39] ROK arms sales reached US$600 million in 2013, according to the authoritative IHS Janes consulting group, and the firm estimated that South Korean arms exports could eventually total around US$1.5 billion per year.[40]

Impediments to increasing defense autarky

South Korea has achieved an undeniably high degree of success in indigenizing arms procurement. The ROK has in a matter of decades built up a domestic defense industry capable of producing a broad range of weapon systems, and as such, approximately 80 percent of its arms are procured indigenously. In many cases the capabilities and quality of these homegrown arms are said to be as high as could be found in the West (as also evidenced by South Korea's recent success in securing considerable overseas arms sales).[41] As a result, it could be reasonably argued that the ROK has attained most of its goals for autarkic military acquisition.

Nevertheless, even after more than 40 years of significant public and private inputs in infrastructure and technology, South Korea still possesses only limited capacities for self-reliant armaments production. Overall, while the country's defense technology and industrial base has "elevated from a third-tier arms producer to a second-tier one" by virtue of considerable effort and investments,[42] much of the local defense sector remains deficient when it comes to innovation and true indigenization. Much of the ROK's so-called "autarky" actually exists only on paper. According to at least one authoritative source, local arms production continues, to a large degree, to rely heavily upon foreign inputs in several critical areas, such as heavy-duty vehicle engines, transmissions, active protection systems (e.g., reactive armor), jet engines, airborne radar systems and other avionics, landing gear, early warning and tracking radar, fire control systems, thermal imagers, laser detection sensors, navigation systems, datalinks, sensor fusion technologies, and signal processing; additionally, the South Korean defense technology base possesses only "limited structural design technologies"; in the field of precision-guided munitions and missile systems its "core technologies (such as seeker design, system optimization, infrared sensors, etc.) remain at a rudimentary level," and surveillance and reconnaissance technologies are "completely new" areas for the local defense industry.[43]

Consequently, South Korea's arms industry is only truly self-sufficient in a few areas, such as small arms, ammunition, and light armored vehicles; in most other cases a considerable proportion of the value of "indigenous production" is still foreign in origin. For example, several critical components found in the K1 main battle tank, such as the engine, transmission, gun, and sight, were originally sourced in the United States and Western Europe (although, to be fair, the more advanced K1A1 tank has achieved a "localization rate" of nearly 83 percent). The

Chunma surface-to-air missile comprises 43 percent of foreign systems by value, mostly French. While local shipyards are constructing the hulls for the KDX destroyer program, foreign firms continue to supply the bulk of its key weapons systems and electronics, particularly its command and fire control systems. The KDX-III, for example, incorporates the US *Aegis* air-defense radar and the Standard (surface-to-air) Missile; overall, only 54 percent of the KDX-III is localized production. Indigenous aircraft programs are particularly dependent on foreign technologies and systems. The KT-1 trainer plane has a localization rate of only 44 percent, while the localization rate for the country's much-vaunted T-50 advanced trainer/light attack jet is only 61 percent.[44] Lockheed Martin partnered with South Korea on the design of the T-50, and as such was responsible for the development of some of the most critical elements of this plane, including the wing, computerized flight-control system, and avionics suite.

Moreover, even after decades of significant public and private investments in infrastructure and technology, much of South Korea's defense technology base remains weak. While the ROK has proven to be very capable at licensed manufacturing and at the indigenous production of such relatively low-tech items as small arms and ordnance, it has experienced much less success when it has come to developing and designing major weapon systems. The K1 tank, for example, was initially deemed to be a mediocre performer (its transmission system has a history of stalling) with poor ergonomics, besides being a systems integration nightmare (due to its diverse collection of US and European components and subsystems); it took years to work out the bugs.[45] The ROK's follow-on indigenous tank, the K2 Black Panther, has experienced similar teething problems. It suffered repeated snags with perfecting its locally developed powerpack (engine and transmission), jointly built by Doosan DST and S&T Dynamics. This eventually led to the temporary suspension of K2 production, and eventually a German-made powerpack was substituted (nevertheless South Korea remains committed to eventually using a locally made engine and transmission).[46] Other domestic weapons systems have suffered similar glitches. The K9 howitzer has experienced problems with software bugs and barrel corrosion; in one instance, when North Korean forces shelled Yeonpyeong Island in November 2010, the ROK Army response was enfeebled because four of six available K9s were inoperable. In addition, on two occasions an ROK Army K21 IFV sank twice during use, killing one soldier; subsequent investigations found critical flaws in the vehicle, including a lack of buoyancy and malfunctioning wave plates and drain pumps.[47]

As in many other countries, "technology overreach" has been the South Korean arms industry's Achilles heel. Early successes with local arms production only bred greater ambitions, which in turn spurred it to pursue programs that increasingly taxed the nation's indigenous technological capabilities to fulfill. While South Korea's defense industry proved to be very adept at licensed manufacturing and the indigenous production of relatively low-tech items such as small arms and ordnance, it experienced much less success when it came to developing and designing major weapon systems. At the same time, the predilection on the part of the central government for high-technology "prestige projects," such as combat aircraft and missile systems, and an overconfidence in the abilities of local industry to quickly move such programs into production tended to result in unexpected development problems, delays, cost overruns, and failure.[48]

Moreover, attempting to undertake such an ambitious range of armaments manufacturing on such a small scale often resulted in highly inefficient and uneconomical structural operations, involving small production runs, high unit costs, and considerable overcapacity in manufacturing. For example, Seoul's insistence on locally assembling the US F-16 fighter added about 20 percent more to the total cost of acquisition than if the ROKAF had procured the aircraft directly from the United States. Moreover, in 2000 the South Korean government decided to procure an additional batch of 20 F-16s – even though the ROKAF itself argued that the planes were unneeded and unnecessary. Rather, the decision was made solely on the basis of keeping to KAI's assembly lines in business until the much-delayed T-50 program entered into production; this stopgap measure cost the ROK US$700 million.[49] Earlier efforts to license-produce the F-5 fighter and MD-500 helicopter resulted in similar cost increases.[50]

Heavy investments in armaments production, especially during its initial stages of defense industrialization, has resulted in considerable excess capacity in the domestic military-industrial complex. During the 1970s, 1980s, and 1990s South Korea greatly expanded its arms-manufacturing capacities in response to existing or projected needs, only to later find itself saddled with overlapping and duplicative capacity and underutilized, high-overhead facilities. Around the turn of the century, for instance, capacity utilization of the overall defense sector was estimated to be at around 60 percent – and only 36 percent in the case of the ordnance and ammunition sectors.[51] Even today it is estimated that, in general, the Korean defense industry operates at just 57 percent of capacity.[52] These diseconomies of scale have greatly undermined the efficiency of local defense firms and have constituted a serious drain on indigenous arms development and procurement.

The presence of excess competition within such a small arms industry only compounded these problems of excess capacity, particularly in the aerospace sector. For example, during the 1990s Korea possessed *four* separate aircraft-manufacturing companies, each of which had invested billions of dollars in new factories and production lines, not only to build the F-16 and the T-50 but also to respond to an ambitious national program to establish the country as one of the world's leading aerospace producers by the turn of the century.[53] These plans included building an entirely indigenous fighter by 2010, as well as a 100-seat regional jet, neither of which came to fruition.[54] Consequently, by the late 1990s the South Korean aviation industry was operating at less than 50 percent of capacity and was at least one billion dollars in debt.[55] Eventually, in 1999 the government forced three of the four aerospace companies to merge into KAI. Even after this amalgamation, however, the domestic aircraft industry remains at overcapacity In particular, the Korean Air Lines Co. continues to engage in various aircraft projects, often in competition with KAI; at one point, in fact, KAL had teamed up with Airbus Industries in (an ultimately unsuccessful) contest to win away the KFX program from KAI.[56]

These problems with hyper-competition are not confined to the aerospace sector. Hyundai Heavy Industries, DSME, and STX Offshore and Shipbuilding all compete aggressively for naval shipbuilding contracts. During the 1990s, for example, DSME was selected to be the sole-source supplier for the Type-209 submarine. However, when the time came to build the Type-214/*Son Wong-il*-class, Hyundai won the bid to produce the ROKN's first three boats in this series, and it built a specialized facility at its Ulsan shipyards for submarine construction. Eventually, DSME received a follow-on contract to build two Type-214s – and STX a contract for a single boat – and consequently all three shipyards are now building Type-214 submarines.

The ROK government has also failed in general to reduce its financial footprint in the domestic arms industry. Seoul, through various state-owned financial institutions, controls a 27 percent stake in KAI and a 50.4 percent share in DSME, the latter of which was acquired when Daewoo went bankrupt in the late 1990s. The central government would like to shed itself of these shares, but recent efforts to divest itself have been unsuccessful, due mainly to the lack of a sufficient number of bidders. At the same time, foreign direct investment in the Korean defense industrial base has also been constrained. For example, aerospace firms, such as BAE Systems, Boeing, Airbus, and Lockheed Martin, have all expressed interest in the past in acquiring a stake in KAI, but the central government has indicated that it would prefer a domestic buyer; in fact, foreign shareholding in KAI is limited to 10 percent.[57]

Overall, therefore, the South Korean arms industries appear to be at a "technology plateau," stuck somewhere in the middle of the military production food chain. The ROK has certainly demonstrated its ability to build very good low- or medium-tech weaponry, such as small arms, artillery systems, light armored vehicles, and ship hulls. It has also proven to be a capable performer when it comes to license-producing foreign weapon systems such as jet fighters or submarines, and a reliable and tech-savvy partner when it comes to joint development programs such as the T-50 (and most likely the KFX program). Nevertheless, weapons manufacturing does not necessarily get any easier the more one strives to develop and manufacture increasingly advanced weapons systems – just the opposite, in fact. Moving on to the next level of armaments production – that is, the indigenous development and production of highly advanced and sophisticated weapons – appears to be exponentially more challenging. If anything, the materiel demands of autarkic advanced arms production – money, R&D efforts, personnel, etc. – may be more than South Korea is presently willing to commit, particularly when such sophisticated weapons can be just as easily purchased from abroad. Even with this commitment, enduring structural challenges such as overcapacity, excess competition, and diseconomies of scale and means that success is not assured.

Conclusion

If the South Korean defense industry has truly reached a plateau, then any more talk of autarky may be moot. If so, what kind of future does the ROK arms industry face? Oddly enough, South Korea remains bullish about its arms industry. The ROK government continues to plow money into new indigenous projects, such as the KFX next-generation combat aircraft, a 3,000-ton submarine, and a homegrown missile defense program. In response, the local defense industry has continued to ramp up its R&D and manufacturing capabilities, pumping out new products such as the K2 tank, the K21 IFV, the T-50 fighter, and the FFX frigate. At the same time, Korean arms manufacturers, with Seoul's blessing and support, have aggressively pursued overseas arms sales, and South Korean weapons exports hit a reported US$600 million in 2013.

In this regard, Korea Aerospace Industries is the embodiment of the country's hopes and dreams for its defense industry. The ROK continues to nurture the ambition of becoming a world-class airframe designer and manufacturer, and it expects KAI to eventually be among the world's leading aerospace-producing countries. The company is also expected to become a major arms exporter. As already noted, it has sold the KT-1 trainer aircraft to Indonesia, Peru, and Turkey, as well as T-50s to Indonesia, Iraq, and the Philippines. Seoul has especially high expectations for its T-50 advanced trainer/lightweight fighter jet; amazingly, KAI expects to export up to 1,000 T-50s over the next decades and to capture one-quarter of the world's market for this kind of aircraft.[58]

In addition, KAI is expanding its commercial aviation activities. It has become a major subcontractor to Boeing and Airbus, for example, supplying parts and components for the Boeing 787 and the Airbus A350XWB, as well as other civil airliner programs. More ambitiously, KAI still hopes to develop and produce its own commercial aircraft. In at least one instance the company was in talks with Bombardier of Canada to collaborate on a 90-seat turboprop passenger plane.[59] Overall, KAI's current vision is to be a "Total Solution Provider" and to rank among the world's top 15 aerospace companies by 2020, with revenues reaching US$10 billion.[60]

In general, South Korea is insufficiently sized to support a full-blown autarkic arms industry. In the first place, its domestic arms market is too small to support efficient production runs, while relying on overseas customers to make up the difference is a risky, failure-prone strategy (it is highly unlikely, for instance, that KAI will export anywhere near to 1,000 T-50 fighters – which is more than 5 times as many T-50s as the ROKAF will buy). Structural problems, such as excess competition and low utilization rates, have only exacerbated these diseconomies of scale. Earlier predictions that KAI would be one of the world's *top ten* aerospace firms by 2005, and then by 2010, came up woefully short.[61] Second, the ROK national technology base is not broad-based enough to support a sophisticated, across-the-board military R&D effort; consequently, there will always exist critical defense dependencies on foreign technologies.

Given all this, technonationalism is critical to explaining why Seoul continues to emphasize the indigenization of armaments production, and in particular why it continues to promote its aerospace sector as a key strategic industry. Rightly or wrongly, South Korea seems confident that it can grow out of its problems, that its defense and aerospace industry can secure sufficient new business so as to avoid any major, painful bloodletting or retrenchment. In particular, Seoul expects to greatly expand its defense exports, even thought it has traditionally been a bit player in the global arms market. Finally, while ambitious plans for several indigenous weapon systems – such as the KFX and a new indigenous submarine – have sometimes been deferred or delayed, they never have been abandoned.

Before one is too dismissive of the prospects for the future of armaments production in South Korea, however, a number of other points must be kept in mind. First, local defense industries are still perceived to be important national assets, and despite its difficulties, the ROK is not likely abandon indigenous arms production or its ambitions to become more self-sufficient in weapons design and development. Such enduring factors as ensuring a secure and reliable source of arms for national defense, aspirations for higher regional status, and even national pride will all continue to drive the expansion of local defense production. In addition, local arms industries are still a center of considerable high-tech activities. The creation of indigenous defense industries has been an important driver of indigenous R&D, as well as a powerful motivation behind the training of research scientists, engineers, technicians, and skilled workers. South Korea has sunk considerable resources, in terms of people, labs, and design and engineering centers, into such areas as military aerospace and defense electronics. Not only do ROK leaders not want to lose these capabilities and talents, but they also hope to continue to spin-off defense know-how and skills to the commercial sector.

Second, the overall technological advancement of South Korea is continuing. The ROK possesses some of the fastest growing high-technology sectors in the world, particularly in the area of electronics, computers and information technologies, telecommunications, and automotive technology. South Korea, for example, is among the world's largest producers of smartphones, tablet computers, semiconductors, integrated circuits, flash memory devices, liquid-crystal displays (LCD), and light-emitting diodes (LED). It is the fourth-largest automaker in the world, and it is still one of the largest and most profitable shipbuilders. Embedding armaments production within such large, advance technology companies, as are most South Korean *chaebol*, provides at least the possibility that breakthroughs in high-tech commercial spheres could percolate over to the defense sector.

Military-industrial planners in South Korea are also paying more attention to the potential military applications of domestic commercially based sources of technology. As the costs and technological demands of defense modernization rise, so too have pressures increased to integrate South Korean military R&D and production more closely with local high-tech civilian industries. In addition, these planners increasingly recognize that domestic commercially oriented industries have lately overtaken indigenous military industrial bases in terms of technology, and that more effort should be given to exploiting such technologies through spin-on. Ideally, this would mean that defense manufacturing would become more embedded in the overall industrial base, with technology flows occurring simultaneously between the military and commercial manufacturing sectors.[62]

Finally, as noted in the introduction, a dependency on foreign military technology is not always a bad thing. For one thing, it can be as a very smart short-term strategy for reducing the high entry costs to advanced arms production; cooperative R&D, in fact, is a critical component of South Korea's long-term technology base development strategy. And given the number of countries (particularly in Europe) prepared to transfer advanced defense technologies to the developing world, recipient vulnerability to supplier controls can be quite low.

In the end, overcoming the technology plateau may be less of a challenge to the ROK than it is for other newly industrialized countries seeking to enter the international armaments business. It possesses a large high-technology base and a highly educated workforce, it has forged close ties with advanced industrialized countries that are willing to transfer those military technologies that South Korea does not yet have, and it has a sizable enough domestic market for indigenously produced arms that, in turn, provides a solid base for oversees sales. Most of all, South Korea's drive to expand its indigenous arms production capabilities is congruent with its rising geopolitical ambitions in the region: this is a critical stimulus that will likely continue to spur the country through its economic and technological challenges to achieve greater autarky.

Notes

1 Jong Chul Choi, "South Korea," in Ravinder Pal Singh, ed., *Arms Procurement Decision Making, Volume I: China, India, Israel, Japan, South Korea and Thailand* (Oxford: Oxford University Press, 1998); p. 183. For more on the ROK defense industry, see Janne E. Nolan, *Military Industry in Taiwan and South Korea* (London: Macmillan, 1986); Janne E. Nolan, "South Korea: Ambitious Client of the United States," in Michael Brzoska and Thomas Ohlson, eds., *Arms Production in the Third World 1971–1985* (Oxford: Oxford University Press, 1987; Kwang-il Baek and Chung-in Moon, "Technological Dependence, Supplier Control and Strategies for Recipient Autonomy: The Case of South Korea," in Kwang-il Baek, Ronald. D. McLaurin, and Chung-in Moon, eds., *The Dilemma of Third World Defense Industries* (Boulder, CO: Westview Press, 1989); Ralph Sanders, *Arms Industries: New Suppliers and Regional Security* (Washington, DC: NDU Press, 1990), pp. 77–88; Richard A. Bitzinger, "South Korea's Defense Industry at the Crossroads," *Korean Journal of Defense Analysis*, Vol. 7, No. 1 (Summer, 1995); Dean Cheng and Michael W. Chinworth, "The Teeth of the Little Tigers: Offsets, Defense Production and Economic Development in South Korea and Taiwan," in Stephen Martin, ed., *The Economics of Offsets: Defense Procurement and Countertrade* (London: Harwood, 1996); Kongdan Oh, "U.S.-Korea Aerospace Collaboration and the Korean Fighter Project," in Pia Christina Wood and David S. Sorenson, eds., *International Military Aerospace Collaboration: Case Studies in Domestic and International Politics* (Aldershot: Ashgate, 1999); Myeong-chin Cho, *Restructuring of Korea's Defense Aerospace Industry: Challenges and Opportunities?* (Bonn: Bonn International Center for Conversion, 2003); Richard A. Bitzinger and Mikyoung Kim, "Why Do Small States Produce Arms? The Case of South Korea," *Korean Journal of Defense Analysis*, Vol. 17, No. 2 (Fall 2005); Chung-in Moon and Jin-young Lee, "The Revolution in Military Affairs and the Defense Industry in South Korea," *Security Challenges*, Vol. 4, No. 4 (Summer 2008), pp. 117–134.

2 Noh, Hoon, *South Korea's "Cooperative Self-Reliant Defense": Goals and Directions*, KIDA Paper No. 10 (Seoul: Korea Institute for Defense Analyses, April 2005), p. 5.
3 Lee Jong-sup, "The ROK-US Alliance and Self-Reliant Defense in the ROK," in Alexandre Y. Mansourov, ed., *A Turning Point: Democratic Consolidation in the ROK and Strategic Readjustment in the US-ROK Alliance* (Honolulu: Asia-Pacific Center for Security Studies, 2005), pp. 246–266; Han Yong-sup, "Analyzing South Korea's Defense Reform 2020," *Korean Journal of Defense Analysis*, Vol. 18, No. 1 (Spring 2008), pp. 129–132; Mikyoung Kim, "The US Military Transformation and Its Implications for the ROK-US Alliance," *IFANS Review*, Vol. 13, No. 1 (July 2005), pp. 21–22.
4 ROK Ministry of National Defense, *Defense White Paper 1999: Republic of Korea* (Seoul: Korea Institute for Defense Analyses, 1999), p. 145.
5 Choi, "South Korea," p. 185.
6 Cheng and Chinworth, "The Teeth of the Little Tigers," p. 269; Oh, "U.S.-Korea Aerospace Collaboration and the Korean Fighter Project," p. 39.
7 "Security Aide Explains Vision of Balance Role," *Joongang Ilbo*, April 13, 2005; "'Balancing Role' Focuses on Possible China-Japan Conflict," *The Korea Times*, March 31, 2005.
8 Chang May Choon, "S. Korea's Rise as a Middle Power," *Straits Times* (Singapore), September 28, 2015.
9 "South Korea's Transforming Alliance with U.S. to Play Neutral Role in Northeast Asia," *The Associated Press*, April 11, 2005.
10 Oh, "U.S.-Korea Aerospace Collaboration and the Korean Fighter Project," p. 39.
11 Bitzinger and Kim, "Why Do Small States Produce Arms?" p. 201.
12 Bitzinger and Kim, "Why Do Small States Produce Arms?" pp. 199–200.
13 Bitzinger and Kim, "Why Do Small States Produce Arms?" p. 199.
14 Nolan, "South Korea: Ambitious Client of the United States," pp. 222–225.
15 Nolan, "South Korea: Ambitious Client of the United States," pp. 218–219; Choi, "South Korea," p. 191.
16 Cheng and Chinworth, "The Teeth of the Little Tigers," p. 249; Choi, "South Korea," p. 199; Robert Karniol, "South Korean Industry: Learning Curve," *Jane's Defense Weekly*, October 22, 2003.
17 Kaan Korkmaz and John Rydqvist, *The Republic of Korea: A Defense and Security Primer* (Stockholm: Swedish Defense Research Agency, 2012), p. 77.
18 Baek and Moon, "Technological Dependence, Supplier Control," pp. 158–159; US Congress, Office of Technology Assessment (OTA), *Global Arms Trade: Commerce in Advanced Military Technology and Weapons* (Washington, DC: Government Printing Office, June 1991), p. 131.
19 *Agency for Defense Development* (Daejeon: Agency for Defense Development, 2008), p. 5.
20 Jon Grevatt, "Peninsular Procurement," *Jane's Defense Weekly*, October 17, 2013; Oh Seok-min, "S. Korea to Raise Defense Spending by 2020," *Yonhap News Agency*, April 4, 2015 (http://english.yonhapnews.co.kr/search1/2603000000.html?cid=AEN20150417009700315).
21 Wesley Spreen, *International Cooperation in the Aerospace Industry* (Kuala Lumpur: ADPR Consult, 1998), p. 118.
22 Jung Sung-ki, "S. Korea Sails Toward Blue-Water Force," *Defense News*, May 28, 2007.
23 Jung Sung-ki, "S. Korea Unveils Cruise, Ballistic Missiles," *Defense News*, April 23, 2014.
24 "LP-X Dokdo (Landing Platform Experimental) Amphibious Ship," *GlobalSecurity.org* (http://www.globalsecurity.org/military/world/rok/lp-x.htm).

25 Jung Sung-ki, "S. Korea Sails Toward Blue-Water Force."
26 "South Korea to Order 5 More U-214 AIP Submarines to Bridge to Indigenous Boats," *Defense Industry Daily*, May 8, 2015 (http://www.defenseindustrydaily.com/KSS-II-South-Korea-Orders-6-More-U-214-AIP-Submarines-05242); Jung, "S. Korea Unveils Cruise, Ballistic Missiles."
27 Jung, "S. Korea Unveils Cruise, Ballistic Missiles."
28 Grevatt, "Peninsular Procurement"; Choi, "South Korea," p. 185.
29 Jung Sung-ki, "S. Korea Aims to Build Indigenous Missile Shield," *Defense News*, March 26, 2012.
30 Tae-jung Kang, "South Korea: Asia's new Powerhouse Arms Exporter," *The Diplomat*, May 25, 2014; Grevatt, "Peninsular Procurement.
31 Jung Sung-Ki, "Hanwha Emerges as South Korea's Defense Giant," *Defense News*, July 29, 2015.
32 Hanwha, which started out in the early 1950s manufacturing gunpowder and explosives, has through its recent acquisition of Samsung Techwin (which includes shares in KAI and the former Samsung Techwin) emerged to be become South Korea's fourth-largest defense company. It earned approximately US$1 billion in defense revenues in 2013, and with the new acquisitions should propel its military sales to around US$2.5 billion per annum. Jung Sung-ki, "With Samsung Sale, Hanwha is S.Korea's New 'Defense Giant,'" *Defense News*, November 30, 2014.
33 Jung Sung-ki, "S. Korea Weighs Designs for KF-X," *Defense News*, February 10, 2014; "S. Korea opts for KAI-Lockheed in $7.8 bn fighter deal," I, March 30, 2015; Dan Darling, "To Little Surprise, South Korea Selects KAI-Lockheed Team to Develop Its KF-X Indigenous Fighter," *Forecast International*, March 31, 2015.
34 Korkmaz and Rydqvist, *The Republic of Korea: A Defense and Security Primer*, pp. 74–75.
35 KAI Website (http://www.koreaaero.com/english/pr_center/cpr_view.asp).
36 SIPRI Arms Trade Database.
37 See Table 1, "Statistics: ROK's Defense Articles Exports," in Richard Weitz, ed., *South Korea's Defense Industry: Increasing Domestic Capabilities and Global Opportunities* (Washington, DC: Korea Economic Institute, November 7, 2013), p. 8.
38 Jung Sung-ki, "S. Korean Arms Industry Emerges as Global Power," *Defense News*, July 16, 2007.
39 KAI website (http://www.koreaaero.com/english/pr_center/cpr_view.asp).
40 Kang, "South Korea: Asia's New Powerhouse Arms Exporter."
41 Baek and Moon, "Technological Dependence, Supplier Control," pp. 164–165.
42 Chung-in Moon and Jae-Ok Paek, "Defense Innovation and Industrialization in South Korea – Assessments, Institutional Arrangements, and Comparative Implications" (paper produced for the Conference on China's Defense and Dual-Use Science, Technology, and Industrial Base, sponsored by the Institute on Global Conflict and Cooperation, University of California, San Diego, July 1–2, 2010), p. 1.
43 Moon and Paek, *Defense Innovation and Industrialization in South Korea*, pp. 8, 12, 14.
44 Moon and Paek, *Defense Innovation and Industrialization in South Korea*, p. 6.
45 Steve Glain, "Seoul Dallies under US Umbrella," *Asian Wall Street Journal*, April 19, 1994.
46 Jung Sung-ki, "S. Korean Vehicles Suffer Gliches, Budget Cuts," *Defense News*, July 18, 2011; Lee Tae-hoon, "Korea to Buy German Engines for K2," *The Korea Times*, April 2, 2012.
47 Jung, "S. Korean Vehicles Suffer Gliches."
48 Nolan, "South Korea: Ambitious Client of the United States," p. 64; Cheng and Chinworth, "The Teeth of the Little Tigers," p. 276.

49 "Air Force Protests Decision to Produce Older Jet Fighters," *Korea Times*, May 12, 1999; Geoffrey Thomas, "Koreans Face Dilemma on Fighter Production," *Aviation Week & Space Technology*, April 12, 1999, p. 71; "S. Korean Air Force Considers Stopgap Fighter," *Flight International*, August 26, 1999, p. 16.

50 Dong Joon Hwang, "Economic Interdependence and Its Impact on National Security: Defense Industry Cooperation and Technology Transfer" (paper presented to the National Defense University Pacific Symposium, Washington, DC, February 27–28, 1992), pp. 12–14.

51 See Cheng and Chinworth, "The Teeth of the Little Tigers," p. 250; Choi, "South Korea," p. 201.

52 Jung, "S. Korean Arms Industry Emerges as Global Power."

53 For example, according to Myeong-chin Cho, Hyundai "did not want to remain as a second-tier supplier, simply manufacturing parts and subassemblies," but rather "aimed at becoming an aircraft system integrator from the beginning." One of its most ambitious efforts – and ultimately costly failures – was to build the wings for the Boeing 717 commercial jet. Hyundai was unable to produce the wing in a cost-effective manner and reportedly lost US$1 million on every wing it produced. Cho, *Restructuring of Korea's Defense Aerospace Industry*, pp. 40–42.

54 At one point during the 1990s South Korea had plans to develop two indigenous aircraft programs, designated the FXX and the FXXX. The FXX was to be an advanced technology multirole fighter, based on existing Western aircraft such as the F-16 or the F/A-18. This project would be followed by the FXXX, a fighter jet based on an entirely indigenous South Korean design; this program was scheduled to start by the middle of first decade of the 21st century. See "FX Fighter Program to Set Stage for Air Force Modernization Plan," *Aviation Week & Space Technology*, June 12, 1989, pp. 191–199.

55 Bruce Dorminey, "Industry Watches as Korea consolidates," *Aviation Week & Space Technology*, November 2, 1998; Bruce Dorminey, "Government Spurns Korean Business Plan," *Aviation Week & Space Technology*, December 14, 1998, pp. 20–31.

56 "For New Jet, It's a Dogfight between KAI and KAL," *Korea Joongang Daily*, February 6, 2015.

57 Grevatt, "Peninsular Procurement."

58 Ju-ming Park, "South Korea Targets Growing Global Defense Market," *Reuters*, August 25, 2013; Joo-hee Lee, "KAI Aims to Export 800 Aircraft by 2030," *Korea Herald*, November 25, 2002.

59 Kyong-ae Choi, "South Korea in Talks to Develop Passenger Plane," *Wall Street Journal*, October 8, 2012.

60 KAI website.

61 Trefor Moss, "South Korean Industry: Industrial Ambition," *Jane's Defense Weekly*, April 11, 2012, p. 32.

62 Weitz, *South Korea's Defense Industry*, p. 3; Chung-in Moon and Jae-ok Paek, *Defense Innovation and Industrialization in South Korea*, SITC Research Brief No. 14, September 2010, Institute on Global Conflict and Cooperation, University of California, San Diego, pp. 4–5; Byun Moo-keun, "Defense Industry Is New Economic Growth Engine," *Korea Times*, undated (https://www.koreatimes.co.kr/www.common/printpreview.asp?categoryCode=270&newsIdx=39728).

6 Other Asian arms producers
Southeast Asia and Taiwan

In addition to the "big four" arms-manufacturing countries of Asia (China, India, Japan, and South Korea), many other states in the Asia-Pacific region have attempted to manufacture their own arms, with varying degrees of effort and success. Nearly every large country in Southeast Asia produces some kind of arms. During the 1980s, for example, Thailand assembled German-designed trainer jets from imported kits, and it has constructed several small naval vessels over the years; most recently, the state-run firm Bangkok Dock finished the assembly of a British-designed offshore patrol vessel, HTMS *Krabi*. Thailand most consistently produces ammunition for its armed forces, along with mortars, rocket launchers, and 4x4 military vehicles.

Vietnamese defense manufacturing is mostly confined to small-scale shipbuilding (such as patrol vessels and corvettes), most of which are based on foreign models. The country is also entering into licensed-production of Russian-designed missile systems, e.g., the *Yakhont* anti-ship missile and the *Igla* man-portable surface-to-air missile. Vietnam's arms production ambitions were dealt a severe blow in 2010, however, when Vinashin (the country's large state-owned shipbuilding corporation) collapsed under a debt burden of US$4.5 billion.[1]

For the most part, however, just three countries in Southeast Asia – Indonesia, Malaysia, and Singapore – dominate the regional arms-manufacturing sphere. Indonesia has invested considerable resources into establishing strategic enterprises covering aerospace, shipbuilding, and land systems. Malaysia has attempted to leverage its few niche capabilities in defense and aerospace manufacturing to become a player in the global supply chain. Singapore – whose indigenous defense industry benefitted particularly from relatively high levels of defense spending over several decades – has developed some highly sophisticated arms-manufacturing capabilities in a few key areas, particularly small arms, armored vehicles, and artillery systems, as well as considerable maintenance, overhaul, and repair (MRO) capacities.

For its part, Taiwan (also known as the Republic of China, or ROC), for a small, newly industrialized economy, has also made some impressive strides in indigenous arms production. Taiwan established its first defense industries in the 1960s. Local armaments production, however, took on an increased importance and sophistication as Taipei's diplomatic isolation grew and its worst fears about being cut off from its principal overseas arms suppliers gradually became a reality. According to Janne Nolan, the Sino-American rapprochement, dating from the Shanghai Communiqué of 1972, particularly prompted the Taiwanese to begin pursuing a strategy of "self-reliance" in arms acquisitions.[2] Initially, the Taiwanese defense industry focused on small arms, such as licensed-production of US M14 and M16 assault rifles, and on providing logistics and maintenance, but local arms-manufacturing efforts both accelerated and expanded in terms of ambition during the 1970s, 1980s, and 1990s. As Nolan put it, Taiwan "deliberately skewed its overall industrialization strategy toward technology-intensive industries" in order to "stay a few jumps ahead" of China, its main rival and security threat.[3] Consequently, Taiwan is one of the few smaller countries in the region to emphasize the indigenous production of tactical missile systems and to successfully design, develop, and manufacture its own advanced combat aircraft, the F-CK-1 Indigenous Defensive Fighter.

Nevertheless, these countries face considerable impediments when it comes to indigenous armaments production, greater even than those confronting the larger arms-producing states of the Asia-Pacific. Some – especially Indonesia – are driven by technonationalist impulses nearly as a powerful as those of the larger arms producers, but they either have already realized – or, in some cases (such as Indonesia and Taiwan) had that realization sadly brought home to them – that autarky is not a viable option. The most that these countries can hope to achieve, therefore, is some kind of "limited autarky," that is, the ability to achieve at least some self-reliance in certain areas of armaments production. In this case they either choose (or are forced to accept) a situation of being *niche* arms producers. In other words, either out of necessity or design they have decided to specialize in certain areas of armaments production and subsequently have jettisoned – or never pursued – other areas of arms manufacturing. This "half a loaf" defense-industrial strategy in part serves as a hedge against foreign suppliers cutting off arms supplies. It also provides states with at least *some* indigenous technological-industrial capabilities to develop customized solutions for national defense. Above all, however, possessing even a modest indigenous arms-manufacturing capability can be an powerful reassurance strategy, encouraging a sense of empowerment and self-confidence in that it can domestically produce at least *something* that contributes to its national security. In this regard especially, limited autarky helps fulfil technonationalist goals.

Indonesia

Indonesia began armaments production in earnest in the mid-1970s with the establishment of several state-owned "strategic enterprises", the most important of which were PT *Industri Pewsawat Terban Nusantara* or IPTN (aviation and aerospace), PT PAL (shipbuilding), and PT Pindad (small arms and munitions).[4] Of these, IPTN was particularly revealing of Indonesia's greater goals for harnessing armaments production for industrial modernization and development. IPTN was the personal brainchild of its founder and first director, B.J. Habibie, who was also the country's Minister for Research and Technology from 1978 to

1998 and who eventually replaced Suharto as president in 1998. Habibie – who also headed up PT PAL and PT Pindad – explicitly viewed the establishment of an aerospace industry as both an instrument and a model for advancing the country's overall technology and industrial base.[5] IPTN in particular was to serve as an indicator of Indonesia's intentions to become a modern industrialized nation and "to prove that a Third World, Muslim-majority country could make a hi-tech leap into global aviation".[6]

The key to realizing these goals was an evolutionary industrial development strategy that explicitly used defense offsets to acquire the necessary research, design, and manufacturing expertise, in order to give "optimal results in the efforts of mastering aviation technology in a relatively short period of time, [i.e.,] 20 years".[7] Offsets are industrial-technological agreements whereby, as part of an arms purchase, the supplier consents to giving back to the purchaser a certain percentage value of the deal (i.e., to "offset" the cost of acquisition), usually in the form of licensed-production rights, technology transfers, or other types of industrial participation in the said weapons acquisition program. Subsequently, IPTN pursued a technology-transfer philosophy dubbed "Begin at the End and End at the Beginning":

> It is a philosophy to absorb advanced technology progressively and gradually in an integral process and based on Indonesia's objective needs. Through this philosophy was thoroughly mastered, not merely materially but also the capability and expertise [of aircraft production]. . . . The philosophy teaches that in building aircraft, it does not necessarily start from components, but [from directly learning] the end of a process (already built aircraft), and then reverse through phases of component manufacturing.[8]

This technology development program was to run through four distinct and progressive phases. Phase one involved the "mastery of manufacturing capabilities" through subcontracting and licensed-production of foreign aircraft designs, "providing the opportunity for both management and the workforce to gain knowledge, skills, and experience".[9] Phase two entailed the integration of technology and the expansion of workforce skills via joint projects with foreign partners. Phase three – technology development – entailed initial efforts to design, develop, and manufacture an aircraft entirely on its own. Finally, during phase four – "basic research" – IPTN was to engage in the indigenous research and development of basic technologies, such as materials, propulsion, and electronics. Breakthroughs made at this stage would be fed into future aerospace programs.

Consequently, IPTN made its start with the licensed-production of several different kinds of foreign-designed aircraft, including the Spanish NC-212 light transport plane, the French NAS-332 helicopter, and the US Bell 412 helicopter. Initial manufacturing entailed the relatively simple assembly of imported knockdown kits, but eventually aircraft in these production runs were to involve greater domestic content until most platforms were built almost entirely indigenously.[10] During this phase IPTN also manufactured components for F-16 fighters and British Hawk trainers purchased by the Indonesia Air Force. In the early 1980s IPTN progressed to the codevelopment of the CN-235 transport aircraft, a joint venture with CASA of Spain. The CN-235 was adapted both for military purposes (as a cargo and maritime patrol aircraft) and for civilian use (as a commuter plane). IPTN has exported the CN-235 to the UAE, Brunei, Malaysia, Pakistan, South Korea, and Thailand. During the 1990s IPTN developed the N-250, a 50-seat turboprop commuter aircraft designed and manufactured entirely in Indonesia (although still using a large number of foreign components, such as the engine, avionics, and landing gear). Two N-250 prototypes were built, and the plane first flew in 1995.

Shipbuilding and small arms followed paths only slightly different from that of IPTN. PT PAL has constructed German-designed 57-meter patrol boats for the Indonesia Navy (TNI-AL), and may build two Dutch-designed *Sigma*-class corvettes. PT PAL also designed a 35-meter patrol vessel and a 60-meter patrol vessel for possible production for the TNI-AL.[11] For its part, PT Pindad produces under license the Belgian FN FNC (SS1 and SS2 series) assault rifle, as well as submachine guns from Italy, mortars from Finland and Israel, and grenade launchers from Singapore.[12] PT Pindad also manufactured sidearms, small arms ammunition, and mortars (but, oddly enough, *not* large-caliber ammunition, such as 105mm or 155mm artillery rounds). Finally, Pindad designed an indigenous 6x6 wheeled armored personnel carrier (APC), modeled after French-made VAB APC and designated the *Panser*.[13]

By the mid-1990s Indonesia's defense industry – and certainly its aerospace business – appeared to be riding high. IPTN in particular enjoyed preferential treatment as a strategic enterprise backed up by strong central government support (Jakarta invested billions of dollars in the company, while Habibie was part of Suharto's inner circle) and a captive market (both the Indonesia military and domestic commuter airlines were often compelled to buy IPTN products).[14] By 1997 IPTN had grown to a workforce of almost 16,000, and the company was intending to become " 'the Toyota of aerospace,' with an aircraft to meet every niche in the 20- to 130-seat range".[15] Its main factory at Bandung featured a state-of-the-art manufacturing facility, including several dozens of advanced computerized numerically controlled tools.[16] It had one major indigenous program (the N-250 commuter plane) already flying and another (the notional N-2130 100-passenger regional jet) on the drawing board. Indonesia appeared to be making considerable strides towards meeting its goals of self-sufficiency and creating a world-class defense/commercial aerospace sector.

Much of this apparent success was illusory, however. In fact, Indonesia was finding it increasingly difficult to break into civil aviation manufacturing at the level of a systems integrator (i.e., the ability to bring together numerous subsystems and components into a single finished product and to have it function as a whole) or even as a major partner in collaborative aircraft programs with foreign firms. The Indonesian government poured nearly $1 billion into the N-250 program, but despite this huge investment, the aircraft continued to experience considerable teething problems.[17] In addition, the plane failed to receive certification from the US Federal Aviation Authority, which made it almost impossible to market the aircraft overseas. At the same time, orders began to dry up for its other products.[18]

The 1997–98 Asian financial crisis was a defining event for Indonesia's defense industry. Jakarta was forced to re-examine and ultimately dramatically scale back its ambitious plans for its aerospace industry, and downsize its arms industry. As a condition of the bailout by the International Monetary Fund, the central government was forced to cut off all support to IPTN, and by 2000 the company had run up a debt of over $500 million. In response, the company underwent a major restructuring, which included a name change (to PT *Dirgantara Indonesia* [PT DI], or Indonesian Aerospace), the divestiture of unneeded production capabilities (particularly in the area of engineering), and, most significant of all, the elimination of more than three-quarters of its overall workforce.[19]

By the end of the first decade of the 21st century, much of Indonesia's defense industry was at a standstill. Most military programs were far behind schedule due to governmental and corporate financial constraints, and actual defense production was nearly non-existent (small arms and ammunition is perhaps the sole exception). The country's showcase company PT *Dirgantara Indonesia* suffered particularly: PT DI was forced to eventually reduce its workforce to around 3,500 employees (which is probably still too many personnel for what little work it currently has), and to cancel several key manufacturing projects, most notably the N-250 and N-2130 commercial airliner projects. For its part, PT PAL has had to lay off 900 of its 2,300 employees. Altogether, Indonesia's defense industrial workforce has shrunk to around 9,000 workers.[20]

Malaysia

The Malaysian defense industry has even less coherence – structurally, organizationally, or conceptually – than does Indonesia's. Malaysia manufactures arms for both military and economic reasons, i.e., to achieve self-reliance in spares and logistics support, modification, upgrades, retrofits; to create high-tech employment, value-added work, and backward linkages in support of local industry (especially small- and medium-sized enterprises, or SMEs); and to obtain high-end sensitive military technology and know-how as well as economic spin-offs to Malaysia's non-defense industrial sectors. In terms of its strategic function, local defense production serves mainly to cater to the needs of the Malaysian Armed Forces (MAF), in particular by helping to sustain equipment purchased from overseas (e.g., through-life support and production of spares and replacement equipment).[21]

Nevertheless, while Malaysia categorizes the defense sector as a "strategic industry", there has actually been minimal central planning for a defense-manufacturing sector, and consequently very little real defense industrialization.[22] The local defense industrial base is a hodgepodge of private and state-owned companies, most engaged mainly in maintenance, repair, and overhaul (MRO) activities or in low-end manufacturing of parts, components, or sub-assemblies. Key arms-manufacturing companies include AIROD (which mostly carries out MRO for Royal Malaysian Air Force [RMAF] aircraft, as well as modifications and upgrades of RMAF airframes; for example, AIROD undertook the conversion of C-130 cargo planes into air-to-air refueling tankers, in addition to the modernization of avionics on imported Mi-171 helicopters); SME Aerospace (which manufactured pylons for Hawk trainer jets supplied to RMAF); SME Ordnance (which undertook licensed manufacture of the US M4 carbine, as well as ammunition, mortars, and explosives); Sapura defense (which produces communications gear and electronic warfare systems); Deftech (a licensed-producer of the Turkish-designed ADNAN tracked armored vehicle); and CTRM (which manufactures the Eagle unmanned reconnaissance aircraft). In addition, the Malaysian government operates the Boustead Naval Shipyard, which constructed the *Kedah*-class Next Generation Patrol Vessel (based on the German MEKO A-100 design) for the Royal Malaysian Navy (RMN).

At the same time, the development of Malaysia's arms industry has been hampered by a number of challenges, which include, according to one former local defense industry executive, "the lack of critical mass [and] high capital investment, lack of research and development, rigid specifications, lack of competencies, absence of uniformity, and lack of promotional and marketing activities".[23] Additionally, the Malaysian defense industry has suffered from an inability to garner any significant overseas arms sales that help create economies of scale (and therefore reduce unit costs) and generate additional income.

However, it is probably the lack of an organized and centrally managed defense research and development (R&D) base that is the most serious impediment afflicting the Malaysian defense sector. Most military-related R&D is self-financed by individual firms, consequently resulting in a series of *ad hoc*, under-financed defense projects (e.g., Deftech's effort to develop an indigenous wheeled armored vehicle).

As with Indonesia, one critical way in which Malaysia has tried to overcome its deficiencies in indigenous R&D capabilities is through offsets, particularly licensed-production and technology transfers; even then this strategy has borne only limited success, mainly due to the still-limited capacity of local industry to absorb and exploit imported know-how.[24] For example, Malaysia is keen to coproduce two *Improved Lekiu*-class frigates being acquired by the RMN; however, BAE Systems, the British-based lead contractor for this program, has been reluctant to include local shipbuilders in any kind of significant industrial cooperation, arguing that they are "not advanced enough" to play a large role in the program.[25]

Poor program oversight has also tended to hinder local military projects, most notably in the *Kedah*-class shipbuilding program. Initially an ambitious plan to build 27 large OPVs, the *Kedah* project was plagued by mismanagement, fiscal irregularities, quality control problems, and delivery delays. The first ship in the series actually failed to pass its pre-delivery sea trials due to technical problems and quality issues, and the resulting scandal forced the Malaysian government to put a new management team in place and to transfer construction to a different shipyard. Eventually the *Kedah*-class program was reduced to just six ships.

Malaysia's current strategy for its indigenous defense sector appears to be a dual-use approach, i.e., spinning off industrial capacity to commercial use or exploiting commercial technologies for military purposes. In particular, Kuala Lumpur is seeking to support the civilian side of Malaysia's defense-relevant industry in terms of building up job skills and technological development.[26] For example, military MRO provider AIROD is now heavily involved in maintenance and overhaul work on several types of commercial aircraft, while SME Aerospace produces parts and sub-assemblies for Boeing, Airbus, and Eurocopter.

None of these initiatives, however, have so far done much to directly increase the capabilities of local arms producers to design, develop, and manufacture indigenous weapons systems, especially more advanced systems. Overall, the Malaysian defense sector still operates on a more or less *ad hoc* basis, emphasizing immediate economic returns with little centralized thought or effort given to ensuring that arms production is sustainable, expandable, or evolutionary, i.e., rising up the technological ladder.

Singapore

The Singaporean arms industry is inexorably linked to the country's concept of "total defense," that is, the idea that the entire resources of the nation must, if necessary, be mobilizable for the sake of national defense. Consequently, the arms industry is an "integral part" of national security.[27] At the same time, Singapore is hardly self-reliant when it comes to military equipment; if anything, the country is heavily dependent on arms imports, particularly big-ticket items such as combat aircraft, helicopters, main battle tanks, submarines, and missile systems of all types. Nevertheless, the maintenance of at least some degree of indigenous armaments production is crucial to the physical and psychological defense of the nation.

As a small nation with limited natural resources, a declining birth rate, a shortage of skilled manpower, no strategic depth, and sandwiched between two large, potentially threatening neighbors, Singaporean armaments production has been focused first and foremost on meeting the immediate needs of the Singapore Armed Forces (SAF). In addition, since high technology is seen as a critical force multiplier, arms procurement decisions have been generally measured against what the Singaporeans can affordably do by themselves and what makes more sense to buy from foreign sources.[28]

When it comes to indigenous arms production and defense industrialization, therefore, Singapore – as opposed to Indonesia, for example – regards any potential economic benefits as secondary to the task of bolstering the country's defense capabilities. As one observer stated, "The Singapore defense industries are not viewed as part of the country's economic development strategy, as they are in Indonesia, but rather are viewed as an integral part of the country's concept of Total defense."[29] Consequently, Singapore has traditionally tended to take a more pragmatic and therefore more selective approach towards defense industrialization. It has never sought nor even harbored the goal of self-sufficiency in armaments production, and the country has remained entirely dependent upon foreign sources for critical weapons systems such as fighter aircraft, helicopters, submarines, tanks, and all kinds of tactical missile systems (air-to-air, surface-to-air, anti-tank, etc.). Instead, the local defense industrial base is geared primarily towards guaranteeing the supply and maintenance of critical systems and towards developing the capability to upgrade and modify imported weapons systems.[30]

Local arms production is centered mainly on the state-owned ST Engineering (STEngg), formerly Singapore Technologies Engineering. STEngg has its roots in Chartered Industries, established in the mid-1960s to produce small arms ammunition for the SAF. After going through several expansions and reorganizations, STEngg presently comprises four main subsidiaries: ST Aerospace (aircraft manufacturing and maintenance); ST Electronics (communications, sensors, software, and combat systems); ST Kinetics (land systems and ordnance); and ST Marine (shipbuilding). By 2014, STEngg employed over 25,000 workers worldwide and boasted revenues of US$4.5 billion.[31]

Singapore has been the most successful in the domestic development and production of small arms, artillery systems, light armored vehicles and certain classes of naval vessels. ST Kinetics has outfitted the SAF with the indigenously designed and manufactured SAR 21 assault rifle, and it also produces machine guns, grenade-launchers, mortar systems, and towed and self-propelled 155mm howitzers. The company also manufactures three light armored vehicles for the SAF and for export: the Bionix tracked infantry fighting vehicle, the Terrex 8x8 wheeled armored personnel carrier, and the Bronco all-terrain tracked carrier. Additionally, during the 1990s the country designed and constructed several *Fearless*-class OPVs (which will subsequently be replaced by another indigenous model), along with four 8,500-ton *Endurance*-class landing ships. Much of the rest of local arms manufacturing has revolved around licensed-production or the maintenance and upgrade of SAF equipment. Licensed-production has generally been employed in those defense sectors where Singapore believes it can best contribute to eventual self-sufficiency in design, development, and production and where it believes that indigenous production can meet national defense requirements *and* still be cost effective and technologically world class. Consequently, Singapore has regularly engaged in Industrial Cooperation Programs with foreign suppliers that require technology transfer and training as part of licensed-production arrangements or off-the-shelf buys.[32] These cooperative agreements in turn have mostly been used to expand the local defense industry's maintenance, repair, and upgrade capabilities. During the 1970s and 1980s, for example, ST Marine constructed both 45-meter and 62-meter missile patrol boats under license from Germany's Lürssen, as well as *Landsort*-class minehunters under license from Sweden's Kockums. Building upon these experiences, Singapore was able to design and construct the indigenous *Fearless*-class and *Endurance*-class vessels. ST Marine has also constructed five French-designed *Lafayette*-class frigates (designated the *Formidable* class by the RSN), in part to learn how to maintain and upgrade these vessels over its multi-decade lifespan. During the 1970s and 1980s Singapore licensed-produced the US M-16 assault rifle, as well as manufactured foreign-designed small arms such as the Ultimax 100 and the SAR 80, before graduating to producing an entirely indigenous assault rifle, i.e., the SAR 21.[33]

Singapore's defense industry has also developed considerable expertise in logistics and depot management, in the maintenance and overhaul of aircraft and aircraft engines, and in ship repair.[34] STEngg, for example, has a collaborative arrangement with Pratt & Whitney to overhaul turbine engine blades, which supports it capacities to overhaul RSAF F-16 fighter aircraft.[35] ST Marine has upgraded used submarines acquired from Sweden. In addition, STEngg has built up its systems design, engineering, and integrations skills necessary to undertake modernization and upgrade programs on behalf of the SAF, including refurbishing and refitting the Republic of Singapore Air Force (RSAF) A-4S fighter aircraft with a new engine and new avionics; upgrading RSAF F-5 fighters with a new cockpit avionics suite and radar; modernizing the army's M-113 armored personnel carriers; and retrofitting the navy's patrol boats with Harpoon anti-ship cruise missiles and Israeli-supplied Barak air-defense missiles.[36]

Overall, Singapore appears to have adopted a core competencies/niche production approach to its defense industries. The country has consciously decided to concentrate arms manufacturing in those areas where it believes it has particular key strengths – and also greater potential either to export its products or to find foreign partners – and has either abandoned or declined to enter into those areas where it believes that such armaments production would not be economically viable or technologically competitive. One area, too, where Singapore's defense industry stands out in Southeast Asia is its growing capacity in defense electronics, such as communications systems, sensors, electro-optics (e.g., targeting systems and seekers, lasers, infrared imaging devices, night-vision goggles, etc.), and data management, as well as the capability to integrate these systems in high-performing packages.

Just as important, perhaps, is the fact that since the mid-1990s Singapore has began to commercialize and also to globalize its defense business. STEngg has engaged in a concerted effort to reduce its dependency upon contracting to the SAF. ST Aerospace (STAe) has greatly increased its commercial subcontracting business over the past decade, manufacturing components for Western companies such as Eurocopter and Boeing, and it has also become a global maintenance and overhaul center for commercial aircraft – consequently, roughly half of STAe's revenues come from non-military work.[37] ST Kinetics has also entered into the manufacture of heavy commercial vehicles.

Just as important have been the Singapore defense industry's efforts to increase its international footprint through codevelopment alliances, joint ventures, and transnational mergers and acquisitions. STAe, for example, is a member of the US-led international consortium currently engaged in the development of the F-35 Joint Strike Fighter. It is collaborating with Eurocopter France in the manufacture and marketing of the EC-120 light utility helicopter and with Israel's Elta to upgrade F-5 fighter aircraft for the Turkish and Brazilian air forces.[38] ST Kinetics also teamed with up with American firm Teledyne Brown Engineering in an (ultimately unsuccessful) attempt to sell its Bionix infantry fighting vehicle to the US Army to meet the latter's Interim Armored Brigade concept. Singapore has signed defense technology collaboration agreements with several other arms-producing countries, including Australia, France, Norway, South Africa, and the United Kingdom. In the case of Sweden it has created a bilateral technology development fund to jointly finance cooperative R&D projects; Swedish-Singaporean cooperation has been particularly close in the areas of undersea warfare and defenses against biological or chemical attacks.[39] Finally, Singapore's defense industry has expanded its overseas operations, with STEngg taking a 25 percent stake in the

Irish company Timoney, which produces suspension systems for armored vehicles, and also acquiring the US shipbuilder Halter Marine as a wholly owned subsidiary. Altogether, STEngg has more than 3,300 personnel employed outside Singapore.[40] Overall, Singapore's success as a producer of relatively advanced military systems can be attributed to its "niche competencies" industrial strategy, as well as to the nation's willingness to maintain a relatively high rate of defense spending, particularly when it comes to military R&D.

Finally, Singapore has had modest success as an arms exporter, a tribute perhaps to the quality and capabilities of its indigenous products, especially its small arms. It has sold the SAR 80 assault rifle to Croatia, the SAR 21 to Brunei, and the CIS 50MG 50mm machine to Indonesia (subsequently produced under license). It exported its Ultimax 100 machine gun to Brunei, Croatia, Indonesia, Peru, the Philippines, Slovenia, Thailand, and Zimbabwe. ST Kinetics' 40mm automatic grenade launcher has been sold to Indonesia (where it is produced under license), Italy (again, produced under license), Georgia, Mexico, Morocco, Peru, the Philippines, Sri Lanka, Thailand, and Uruguay. In addition to exporting small arms, Singapore has upgraded the F-5 fighter aircraft for the Turkish and Brazilian air forces and, most recently, closed a deal with Bangkok to supply the Thai navy with an amphibious assaults ship based on the *Endurance*-class design.

Taiwan

Not surprisingly, Taiwan's national defense priorities are geared primarily toward deterring or defending against an attack from mainland China: a naval blockade, an air/missile attack, or even a limited or full-scale invasion. As a result, increased priority has, over the past 20 years, been given to strengthening the ROC Air Force (ROCAF) and ROC Navy (ROCN) for long-range patrolling, screening, and defensive operations, including air interdiction, anti-air and anti-missile defense, anti-submarine warfare, and anti-landing operations.[41]

Armaments production in Taiwan has long traditionally been concentrated in three large, government-owned and -operated enterprises: the Chung Shan Institute of Science and Technology (CSIST), responsible for the development and manufacture of indigenous missile systems; the Aerospace Industry Development Corporation (AIDC), in charge of combat aircraft; and China Shipbuilding Corporation (now known as CSBC Corporation), which constructed warships for the Republic of China Navy (ROCN). In addition, the Ordnance Readiness Development Center of the ROC Ministry of National Defense (MND) is responsible for production of armored vehicles, such as the indigenously developed CM-32 wheeled infantry combat vehicles.

Of these defense enterprises, CSIST is the most important, and certainly the most active at present. CSIST, located just outside Taipei and administered by the Armaments Bureau of the Taiwanese MND, has developed and manufactured all types of anti-ship, air-to-air, and surface-to-air missiles. These include:

* The Hsiung Feng (HF) family of cruise missiles. The HF-I, HF-II, and HF-III are all anti-ship cruise missiles (ASCMs); the HF-III is a ramjet-powered

supersonic ASCM capable of reaching speeds up to Mach 2. CSIST also manufactures a land-attack cruise missile (LACM), designated the HF-IIE, with a range of at least 2,000 kilometers.

- The Tien Chien (TC) series of air-to-air missiles (AAMs). The TC-I infrared-guided short-range AAM is similar in design and missile to the US AIM-9 Sidewinder, while the medium-range Tien Chien II AAM is a fully active radar-homing missile in the same class as the US-made AMRAAM. The TC-I has been operational since the early 1990s, while the TC-II entered production in the late 1990s. In addition, an anti-radiation variant of the TC-II, designated the TC-IIA, was developed to attack ground-based air-defense radars.
- With regard to surface-to-air missiles (SAMs), CSIST has modified the TC-I into a short-range SAM, and has also developed the Tien Kung (TK-I) and TK-II SAM systems for low-to-medium engagements. Finally, it has also developed the TK-III anti-tactical ballistic missiles (ATBM) for defense against ballistic missile threats.

Other CSIST products include the Yun Feng long-range LACM, multiple rocket launchers, unmanned aerial vehicles, and cluster bombs. CSIST's Third Institute is also Taiwan's leading producer of C4ISR systems, radar, fire control systems, missile guidance systems, electro-optical equipment, and electronic warfare systems, while its Chemistry Institute is responsible for developing and producing propellants, warheads, and explosives, as well as chemical warfare defenses.[42]

Taiwan has also put considerable resources into building up a military aircraft industry. In 1969 it created the Aerospace Industry Development Corporation, a government-owned company based in Taichung. During the 1970s AIDC licensed-produced the US F-5 fighters, building over 300 of these aircraft. It also assembled 118 US UH-1 utility helicopters. Starting in the 1980s, however, AIDC began to develop and manufacture its own indigenous aircraft, starting with the TCH-1 primary turboprop trainer in the 1970s and the AT-3 advanced trainer/light attack jet in the 1980s; 50 TCH-1s and 64 AT-3s were built for the ROC Air Force (ROCAF). AIDC's zenith entailed the design, development, and manufacture of the F-CK-1 Indigenous Defensive Fighter (IDF), a lightweight, all-weather supersonic fighter similar in appearance and mission to the F-16. The F-CK-1 is outfitted with a multimode pulse-Doppler radar, and it carries both TC-I and TC-II air-to-air missiles and the Wan Chien cluster munition. The first IDF prototype flew in May 1989 and the aircraft entered series production in 1994; a total of 130 F-CK-1 (both single- and dual-seat) were built for the ROCAF between 1994 and 2000.

CSBC, formerly known as the China Shipbuilding Corporation, with shipyards in Kaohsiung and Keelung, has long been Taiwan's main naval production facility. CSBC has built several types of surface combatants for the ROCN, beginning with the construction of Israeli-designed *Dvora*-class missile patrol boats in the 1970s. During the 1990s it built eight *Chengkung*-class guided missile frigates – a licensed-produced variant of the US Perry-class frigate – equipped with Hsiung Feng ASCMs, surface-to-air missiles, torpedoes, and anti-submarine helicopters.

More recently, CSBC constructed 30 500-ton *Kuang Hua VI*-class missile patrol boats, Taiwan's first indigenously designed warship.[43] Another key shipyard for the ROCN is the Lung Teh Shipbuilding Company, which is currently building 12 *Tuo Chiang*-class high-speed stealth corvettes, featuring a wave-piercing catamaran hull design.

By the turn of the century Taiwan had built up some of the most impressive defense industrial bases among the newly industrialized economies. However, this was about the same time that the tide began to turn against the Taiwanese arms industry. Many domestic weapons programs – particularly the F-CK-1 fighter – were prematurely halted; meanwhile, anticipated successor programs failed to materialize. For example, most of Taiwan's aviation industry was centered around the production of the F-CK-1 fighter for the country's air force; consequently, AIDC built a modern production facility to undertake an expected production run of 250 IDFs. However, in 1992 the ROCAF announced its intention to purchase 150 US F-16s and 60 French Mirage-2000s, and acquisition of the F-CK-1 was subsequently cut back to just 130 planes, the last of which was delivered in January 2000. Consequently, AIDC factories were soon taken up with "idle assembly lines" and "large rooms filled with little-used machines."[44] Other defense sectors faced similar difficulties. CSBC was plagued by low worker productivity and heavy debts, but fierce opposition from labor unions, as well as concerns for the company's long-term economic viability, made it hard to rationalize shipyards.[45] For its part, CSIST, threatened by large numbers of imports of air-to-air missiles (acquired as part of the F-16 and Mirage-2000 buys), at one time planned to greatly curtail its range of R&D activities and to reduce its workforce.[46]

At the same time, many alternate industrial readjustment programs were unsuccessful. Taiwanese defense firms failed to expand arms exports (which were never very large in the first place), while efforts to diversify into commercial ventures proved problematic, to say the least. Taipei had in particular targeted civilian aerospace as a key business sector marked for growth and development, and AIDC has long attempted to expand into civil aviation manufacturing. For example, starting in 1997 it tried to partner with the Czech firm Aero Vodochody to jointly develop and produce the AE-270 small business plane; the plane never won any orders, however, and it was finally cancelled in the late 2000s. One bright spot for AIDC is a current program to upgrade the ROCAF's fleet of F-CK-1 with new radar, avionics, and electronic warfare systems; even then this program will wrap up by 2017, with no new military aircraft programs forthcoming.[47] Meanwhile, schemes to encourage civil-military integration activities involving CSIST's R&D facilities and to promote dual-use technology development also never materialized.[48]

By the early 21st century, therefore, the Taiwanese arms industry was faced with declining orders and excess capacity, both in terms of personnel and production space. At the same time Taipei found it difficult to privatize or rationalize its state-owned defense firms (in terms of workforce reductions) for fear that it could mean the permanent loss of critical high-tech, national security-related industries.

AIDC and CSBC faced a particularly bleak future; only CSIST appeared safe due to its position as a supplier of critical missile systems.

Recent defense industrial reform efforts

Based on this brief overview, one might be justified in being pessimistic about the prospects for local arms industries in Southeast Asian and in Taiwan. Except for certain niches, particularly in Taiwan's missile sector, the overall technological level of *indigenous* military products in these states remains rather low. For the most part the most advanced armaments produced locally are still foreign sourced, that is, either licensed-produced or based on imported designs and technologies. Even Singapore, for all its success in building up the largest arms industry in Southeast Asia and given the consistently high quality of its weaponry, is still only truly self-reliant in the design and development of a handful of land ordnance systems: assault rifles and small arms, ammunition, artillery systems, and light armored vehicles; the rest is all imported or, at most, licensed-built. Indonesia, easily the most ambitious of the Southeast Asian nations in its goal of building up a large and innovative arms industry, has consequently suffered the most embarrassing failures; it has only developed one more-or-less indigenous defense product – the CN-235 transport plane – and even that was in cooperation with CASA of Spain. In addition, the arms-manufacturing capabilities of these industries – with the exception, in certain sectors, of Singapore and Taiwan – is still quite low, concentrated mainly in areas such as the fabrication of parts and components, maintenance and repair, and, in a few cases, final assembly. Consequently, local arms industries remain highly dependent on foreign partners for manufacturing technologies and processes. Even Taiwan's so-called Indigenous Defensive Fighter was highly reliant on foreign inputs: the design and development of its airframe was overseen by Lockheed Martin, the engine is an upgraded version of Allied Signal's TFE-731 turbofan, and the radar was derived from the General Electric APG-67 radar originally developed for the F-20 fighter.

That said, there are indicators that some defense-industrial sectors in these countries may be experiencing a bit of a revival. There is a newfound ambition and effort underway in several states in the region to revitalize and rebuild their ailing defense industries, or to attempt to create new centers of armaments production. Indonesia has been particularly aggressive recently in its efforts to revitalize its defense industry in conjunction with an ambitious effort to expand its armed forces.[49] For example, Jakarta recently decided to acquire up to 16 new patrol boats (outfitted with Chinese anti-ship cruise missiles), which are being manufactured on Batam. In addition, a deal to buy submarines from South Korea will include technology transfers to support the construction of at least one boat in Indonesia. Other prospective new indigenous projects include two types of amphibious assault vessels and a 2,500-ton frigate.[50] Indonesia is attempting to develop a new patrol vessel based on a radical wave-piercing trimaran design, relying heavily upon technology from Sweden (the lead vessel in the series was destroyed in a fire in late 2012, and its current status is uncertain). Most significant of all, in 2011 Jakarta signed an agreement with Seoul to codevelop a new state-of-the-art stealth fighter jet, designated the KFX, which is currently being undertaken by the Koreans; Indonesia could invest as much as $1 billion into the project, for which it would receive a 20 percent stake in the manufacturing program and acquire up to 50 aircraft. If successful, this program would provide considerable stability (both in terms of revenues and employment) for PT *Dirgantara Indonesia* for years to come.[51] To help keep the country's arms industries afloat in the short term, in 2011 the Indonesian parliament approved a 7.8 trillion rupiah (US$700 million) bailout to PT DI, PT PAL, and PT Pindad; in 2012 the government provided PT DI with an additional trillion rupiah (US$90 million) bridge loan to help revitalize the firm.[52]

Malaysia is also planning a significant, if less ambitious, expansion of its arms-manufacturing capacities. The Tenth Malaysia Plan (2011–16) entailed the acquisition of next-generation patrol vessels, multipurpose support ships, and perhaps another squadron of fighter jets. Many of these procurement programs will entail local production or industrial participation. For example, Kuala Lumpur has ordered six new Littoral Combat Ships based on the French *Gowind*-class corvette, which will be built locally by the Boustead Naval Shipyard in Perak.[53]

Particularly when it comes to Southeast Asia, one factor that is particularly favoring local ambitions when it comes to arms manufacturing is the recent rise in regional military expenditures. According to the Stockholm International Peace Research Institute (SIPRI), Malaysia's military budget nearly doubled in real terms between 2000 and 2014, while Indonesian defense spending increased nearly threefold. Consequently, these countries theoretically have more money to spend on indigenous armaments projects.[54]

In addition, Southeast Asian governments have increasingly emphasized offset requirements as a means of gaining new technologies and human capital skills to support more advanced armaments production. In recent years, when they have purchased arms from abroad, Southeast Asian countries have increasingly demanded more (and larger) offsets – such as local licensed-production of the weapon systems being acquired, training and other types of skills building, technology transfers, subcontracts, or foreign direct investments (FDI) in their local arms industries – and in turn attempting to use those offsets to help modernize and reinvigorate their defense sectors.[55] In 2011, for example, Malaysia instituted a new offsets policy that greatly strengthened offsets requirements, in particular creating a target of 100 percent offsets (i.e., offsets equal to the purchase price of the acquisition program), permitting FDI to count toward offset credits, pressuring foreign suppliers to aid in the export of Malaysian defense products, and discontinuing offset multipliers (complicated counting systems whereby suppliers can get a higher relative value for their offsets – for example, a supplier providing advanced training and education to local personnel may be awarded a higher offset value than its actual cost, as it hopefully results in long-term skills development). Additionally, Kuala Lumpur reiterated its commitment to greater oversight and evaluation to ensure that offsets would meet their intended goals.[56]

For its part, Indonesia passed a new defense Industry Law in 2012 that, among other things, addresses the issue of increasing offsets and technology transfers to the national defense industrial base.[57]

For its part, in the second decade of the 21st century Taiwan has finally begun to find some traction in its efforts to commercialize and civilianize its aerospace and shipbuilding. At one time Taipei had planned on denationalizing AIDC by the end of the 1990s and selling at least part of the company to private – and even foreign – investors.[58] In 1996, as a first step toward this goal the government transferred control of AIDC from the Ministry of Defense over to the Ministry of Economic Affairs, and AIDC was subsequently restructured as a state-owned enterprise.[59] Plans were also made to privatize CSBC by late 1997. However, despite missing numerous deadlines for privatization, CSBC was finally denationalized in 2008, and the AIDC in 2014. (Even then, the government still owns large amounts of stock in both companies: a 38 percent share in CSBC, and a 46 percent share in AIDC). AIDC finally turned a profit in 2007, and it has been in the black ever since.[60] At the same time, there are *no* plans to privatize or commercialize CSIST, which remains directly owned and operated by the Ministry of National Defense.

In particular, too, AIDC has been more successful lately in entering into strategic alliances, joint ventures, and subcontracting agreements with foreign aerospace firms. The company is currently collaborating with Sikorsky Aircraft to coproduce cockpits for the S-92 commercial helicopter. It is also working with Western aircraft manufacturers to supply parts and components for commercial aircraft. AIDC is building empennages (tail sections) for the Alenia C-27 transport aircraft and Bombardier's CL-350 business jet, as well parts for the Boeing 787 and the Airbus A320/A321.[61] The company has also expanded its aero-engine business, manufacturing parts and components for foreign aerospace firms, including General Electric, Honeywell, Snecma, and Rolls Royce. Consequently, AIDC has invested millions in building new engine and precision-forging plants.[62] For its part, CSBC has increased its diversification into commercial shipbuilding, particularly container ships, bulk carriers, and oil tankers.[63]

Conclusion

Despite these recent defense budget increases and new offset requirements, it is difficult to be optimistic about the future of arms production in Southeast Asia and Taiwan. Overall, most regional arms industries face a difficult time ahead, and sustaining or expanding their indigenous defense industries will be an ongoing challenge. In particular, few of these countries will likely be able to do much over the next decade to increase their technological capabilities or to expand their defense industrial footprint beyond a few showcase programs. Many of these nations still lack sufficient numbers of qualified engineers and technical personnel to engage in more advanced types of armaments production.[64] It is, for example, difficult to see the Indonesians, given the relatively backward state of their aerospace industry, making much of a contribution to a fifth-generation fighter program. Even Singapore is unlikely to expand its arms production capabilities. About the only area where countries in the region may attain any significant degree of defense-industrial self-sufficiency will remain in the realm of small arms and light weapons (and again, mainly under license from foreign manufacturers).

Despite some recent turnaround, the Taiwanese defense industry has greatly downsized its vision and scope, particularly on the part of AIDC. Except for a limited upgrade of the ROCAF's F-CK-1 fighters, AIDC is currently without a major aircraft-manufacturing program; meanwhile, most of its more ambitious projects for commercial aviation never materialized. Consequently, AIDC has been largely relegated to the role of minor subcontractor to foreign aerospace companies. CSBC was aided temporarily by a contract to build 30 *Kuang Hua VI* missile patrol boats for ROCN, but production of this class of vessel ended in 2011; consequently, CSBC also lacks any current major naval shipbuilding programs. CSIST appears to be Taiwan's only really successful defense enterprise, and it continues to be actively developing and manufacturing many tactical missile systems and other types of military equipment. At the same time, it should be noted that CSIST remains a wholly owned subsidiary of the Ministry of National Defense and is thus shielded from market forces (as well as being opaque when it comes to outside scrutiny and assessment).

Therefore, for most of these countries the barriers to becoming significant arms producers remain high. The capability to manufacture most types of advanced weapons systems – especially highly sophisticated systems like combat aircraft, missile systems, and even surface combatants (hulls are relatively easy to build, but the combat systems that go on them are incredibly complex) – is beyond the reach of most middling or even relatively industrialized countries. Consequently, except for a few sectors where the basic technologies are relatively mature and accessible (such as small arms or (in the case of Singapore) armored vehicles and artillery systems, or (in the case of Taiwan) missile systems), most advanced arms production will remain elusive – and this will most likely continue to be the case no matter how much states raise defense spending or attempt to shortcut the process of armaments development through offsets.

In the first place, while it is true that many countries in the region have experienced significant increases in their defense budgets in recent years, it is unlikely that these spending increases will be sufficient or consistent over a long enough period of time to significantly bolster local arms industries, or that major funding will be channeled into indigenous arms programs in the first place. Indonesia's defense firms (especially PT *Dirgantara Indonesia*) still remain starved of funding, while Malaysia's ambitious military modernization program (new ships, submarines, fighter aircraft, tanks, etc.) may eventually end up outstripping Kuala Lumpur's ability to pay for it all.[65] Defense spending in Southeast Asia, despite growing at an impressive annual rate, is still relatively low (only Singapore barely cracks the $10 billion barrier). Military budgets of a few billion dollars are hardly enough to underwrite robust defense industrial bases.

At the same time, regional experiences with defense offsets could also be taken with a grain of salt. Offsets offer no great shortcuts, either economic or technological, when it comes to achieving viable, self-sustaining defense industries. Arms production is a "capital-and technology-intensive industry" requiring significant investments in equipment and personnel,[66] and offsets alone are insufficient to provide for these requirements, especially as a country attempts to move up the ladder of production. Indigenous – and often quite substantial – sources of financial, industrial, and human capital must also exist independently in order for a nation to make progress towards the independent development and production of advanced weapons systems.[67] To put it another way, offsets cannot substitute for a strong science and technology base.

An observation made nearly two decades ago with regards to the efficacy of offsets in defense industrialization in South Korea and Taiwan is equally appropriate in the case of Southeast Asian arms industries:

> Offsets . . . have had limited impact in fulfilling larger plans for becoming self-sufficient producers across a range of systems (much less becoming global players in high tech industries). This apparent failure could be attributable to overly ambitious plans by central governments, as well as a measure of naiveté in understanding the dimension of domestic resources needed to fully exploit such transfers. There is no doubt that domestic capabilities have grown . . . as a result of technology licensing, production buybacks and other forms of offsets . . . However, offsets have not resulted in anything approaching the creation of global competitors in a vast range of systems . . . nor are they likely to in the immediate future.[68]

Consequently, it is unlikely that Indonesia, Malaysia, Thailand, or Vietnam will ever rise above their current positions as relatively minor players in the global hierarchy of arms producers. Arms production will continue to be *ad hoc* and sporadic, limited to mainly to final assembly work (if even that), low-end manufacturing (parts or sub-assemblies), or a few showcase arms projects that are generally low-tech in nature (small arms, ammunition, small naval vessels, etc.). In particular, the capabilities of the Indonesian or Malaysian defense industry to develop and produce truly transformational technologies related to modern concepts of network-centric warfare (NCW) (which entails the heavy employment of high-technology systems such as unmanned aerial vehicles, precision-guided weapons, and cyber-warfare) will remain very limited, if not nonexistent. Even Singapore, with its comparatively advanced defense industrial base – especially in the area of defense electronics and systems integration – is finding it difficult to engage in a technology-intensive, NCW-based "revolution in military affairs."

In recent years there have been calls to promote greater regional cooperation when it comes to developing and producing arms. In May 2011, for example, at the Fifth ASEAN defense Ministers meeting in Jakarta the participants signed the ASEAN defense Industry Collaboration (ADIC) initiative.[69] ADIC seeks to enhance defense industrial inter-dependence within ASEAN and to foster indigenization and technological independence when it comes to local armaments production. The initiative is focused mainly on initiatives to promote collaborative defense projects, increase regional competitiveness in various dual-use sectors that could have a "spin-on" effect on local armaments production, and in general assist the development and growth of ASEAN's defense industry. ADIC collaboration revolves around education and training in the defense industrial sector, regional partnerships, joint ventures and coproduction, cooperative military R&D, and joint promotion of military equipment in sales and marketing. Despite all these hopes and intentions for ADIC, little has actually come out of it. Most ASEAN governments and militaries remain suspicious about sharing their technologies or loosening protectionist controls over their national defense industries. In addition, given the uneven levels of defense industrial development in the region, many nations are unable to participate as meaningful partners in any collaborative arms program. Consequently, ADIC remains a stillborn effort.[70]

Overall, armaments production in Southeast Asia and Taiwan, despite admirable efforts of late to restructure and revitalize local defense industries, will remain a difficult affair, *even if the goal is "just" limited self-reliance.* Only Singapore will likely be any kind of a player in the global arms marketplace – and even then, it will likely remain only a niche manufacturer in a few, relatively low-tech fields (e.g., small arms, armored vehicles), except perhaps in a few areas such as weapons upgrades and systems integration, where Singapore has some additional expertise. This is not to say that sometime in the future other regional states may eventually break in to the higher reaches of the global arms market, but it will take time and considerable effort before they can do so. They would require not only continued overall industrial and economic development but also a concerted long-term effort to build up their national science and technology bases, an expansion of tertiary education in cutting-edge high-technology fields (such as aeronautical engineering, information technologies, electro-optics, biotechnology, etc.) in order to grow human capital skills, the direction of greater resources into military R&D, a modernization of defense manufacturing capabilities, and above all the maintenance of a consistently high enough rate of military expenditures in order to fund all this activity. Even then, for these nations to be considered "successful" arms producers they will probably need to concentrate on a few core areas where they can carve out a competitive niche for themselves in the global arms market; in other words, attempting to achieve across-the-board capabilities when it comes to arms manufacturing will likely remain forever beyond their grasp in terms of economic sustainability and technological viability (Indonesia, in particular, needs to learn this lesson). This means playing a decidedly narrow or subordinate role in the international arms business, either as a producer of a few high-quality military systems or as a subcontractor to the larger, already established (Western) arms manufacturers. Such an industrial strategy may not be glamorous, but it will likely be more profitable.

Alternatively, of course, these countries could simply decide to remain at the low-tech end of arms manufacturing, e.g., constructing hulls for naval vessels, small arms production, licensed assembly of imported weapons kits, etc. Even then, such a modest approach to arms production will still be a challenge, as it will be continually *ad hoc* and difficult to make economically sustainable. At the same time, it will not make much of a contribution to these nations' national defense efforts, let alone any technonationalist military-technological autarky.

However, before becoming too sanguine about the future of these smaller Asian countries' defense industrial capacities, it is important to keep in mind the extremely high barriers to advanced armaments production that still exist. For these countries, therefore, any technonationalist defense-industrial strategy – even a limited approach – is so arduous and demanding so as to be inconceivable. In some cases – such as the Indonesian and Taiwanese endeavors with aircraft manufacturing – technonationalist goals have become even *more* distant in recent years. Significant challenges – many of them of long standing – still stand in the way of smaller states becoming successful arms producers, as the recent experiences of Indonesia, Malaysia, Singapore, and Taiwan have shown.

Notes

1 "Vietnam's Communist Party Congress Faces Economic Test", *BBC News*, 12 January 2011 (http://www.bbc.com/news/world-asia-pacific-12151218).
2 Janne E. Nolan, *Military Industry in Taiwan and South Korea* (New York: St. Martin's Press, 1986), p. 48.
3 Nolan, *Military Industry in Taiwan and South Korea*, p. 49.
4 *IHS Janes Navigating the Emerging Markets: Indonesia*, op. cit., pp. 20–23; Singh, "ASEAN's Arms Industries", op. cit., p. 251.
5 Huxley and Willett, *Arming East Asia*, op. cit., p. 50; John Bailey, "Habibie's Grand Design", *Flight International*, 19 February 1992, pp. 51–52.
6 Margot Cohen, "New Flight Plan", *Far Eastern Economic Review*, 2 March 2000, p. 45.
7 *Evolution and History of the Indonesia Aviation Industry*, from Indonesian Aerospace's website (http://www.indonesian-aerospace.com/aboutus.php?m=aboutus&t=aboutus8).
8 Ibid.
9 Bailey, "Habibie's Grand Design", op. cit., p. 52.
10 Ibid.
11 Robert Karnoil, "Country Briefing: Indonesia – Interior Designs", *Jane's Defense Weekly*, 21 March 2004.
12 Singh, "ASEAN's Arms Industries", op. cit., p. 251.
13 Karnoil, "Country Briefing: Indonesia – Interior Designs", op. cit.
14 "Airtech CN235," *Jane's All the World's Aircraft* (London: IHS Jane's Information Group) (https://janes.ihs.com/CustomPages/Janes/DisplayPage.aspx?DocType=Reference&ItemId=+++1342564), accessed online October 24, 2013.
15 Bailey, "Habibie's Grand Design", op. cit., p. 52; Cohen, "New Flight Plan", op. cit., p. 46.
16 Andrzej Jeziorski, "Toughing It Out", *Flight International*, 25 July 2000, p. 77.
17 Cohen, "New Flight Plan", op. cit., p. 45.
18 Jeziorski, "Toughing It Out", op. cit., p. 77; Cohen, "New Flight Plan", op. cit., p. 46.
19 Cohen, "New Flight Plan", op. cit., p. 45; *IHS Janes Navigating the Emerging Markets: Indonesia*, op. cit., p. 21.
20 *IHS Janes Navigating the Emerging Markets: Indonesia*, op. cit., p. 21.
21 Kogila Balakrishnan and Ron Matthews, "The Role of Offsets in Malaysian Defense Industrialization," *Defense and Peace Economics* 20, No. 4 (August 2009):

342–345; Kogila Balakrishnan, *Defense Industrialization in Malaysia: Development Challenges and the RMA*, unpublished manuscript, 2008, pp. 4–14.

22 Balakrishnan, *Defense Industrialization in Malaysia*, op. cit., p. 7.

23 Robert Karnoil and Dzirhan Mahadzir, "Country Briefing: Malaysia – The Big Push," *Jane's Defense Weekly*, 24 November 2005.

24 Balakrishnan and Matthews, "The Role of Offsets in Malaysian Defense Industrialization", op. cit., pp. 345–350.

25 Gordon Arthur, "The Malaysian Armed Forces: Celebrating 50 Years of Nationhood", *Asian Military Review*, December 2007, p. 8.

26 Balakrishnan, *Defense Industrialization in Malaysia*, op. cit., p. 9.

27 Bilveer Singh, "ASEAN's Arms Industries: Potential and Limits," *Comparative Strategy*, Vol. 8 (1989), p. 251.

28 Huxley and Willett, *Arming East Asia*, op. cit., p. 50; Tan Peng Yam, "Harnessing Defense Technology: Singapore's Perspective", *DISAM Journal of International Security Assistance Management*, Vol. 21, No. 3 (Spring 1999) (http://www.disam.dsca.mil/Pubs/Indexes/v.21_3/Yam.pdf).

29 Singh, "ASEAN's Arms Industries", op. cit., p. 259.

30 Ron Matthews and Nellie Zhang Yan, "Small Country 'Total Defense': A Case Study of Singapore", *Defense Studies* 7, No. 3 (September 2007): 384.

31 Tim Huxley, "Singapore's Defense Procurement, Research & Development, and Industry", conference paper presented at *Transformation in Global Defense Markets and Industries*, London, 4–5 November 2000, pp. 11–12; Robert Karniol, "Singapore's Defense Industry: Eyes on Expansion", *Jane's Defense Weekly*, 30 April 2003.

32 Matthews and Yan, "Small Country 'Total Defense': A Case Study of Singapore," p. 392.

33 Huxley, *Singapore's Defense Procurement*, op. cit., p. 14.

34 Data derived from ST Engineering's website (http://www.stengg.com).

35 Karniol, "Singapore's Defense Industry", op. cit.

36 Huxley, *Singapore's Defense Procurement*, op. cit., pp. 13–14.

37 Huxley, *Singapore's Defense Procurement*, op. cit., pp. 16–17; Karniol, "Singapore's Defense Industry", op. cit.

38 Huxley, *Singapore's Defense Procurement*, op. cit., p. 19.

39 Ibid., p. 10.

40 Data derived from ST Engineering's website (http://www.stengg.com).

41 David Shambaugh, "Taiwan's Security: Maintaining Deterrence amid Political Accountability," *China Quarterly*, December 1996, pp. 1303–1317.

42 Globalsecurity.org, "Chungshan Institute of Science and Technology" (http://www.globalsecurity.org/wmd/world/taiwan/csist.htm).

43 "Taiwan Launches Patrol Ship," *Jane's Defense Weekly*, September 2, 1998.

44 Brent Hannon, "Changing the Guard: Taiwan's Aerospace Industry is Facing up to the Realities of New Privatization," *Flight International*, March 5, 1997, pp. 49–50.

45 Author's interviews in Taiwan.

46 "Taiwan Institute Cutting Back Sharply Due to Foreign Supply," *The Estimate*, June 10, 1994, p. 4.

47 Defense-update.com, "Taiwan To Upgrade Half F-CK-1 of its Fleet By Year's End," January 3, 2013 (http://defense-update.com/20130103_taiwan-to-upgrade-half-f-1ck-fleet-by-years-end.html#.Vd6dgiWqqko).

48 *1998 National Defense Report: Republic of China*, p. 113, 139.

49 Jon Grevatt, "Briefing: Indonesia Unveils Defense Industry Revitalization Plan", *Jane's Defense Industry*, 11 November 2009.

50 *Revitalizing Indonesia's Defense Industrial Base: Agenda for Future Action*, RSIS Policy Report (Singapore: S.Rajaratnam School of International Studies, 5 July 2012), p. 5.

51 Jon Grevatt, "Indonesia and South Korea Sign KFX Next-Generation Fighter Deal", *Jane's Defense Weekly*, 16 March 2011.
52 *Revitalizing Indonesia's Defense Industrial Base*, op. cit., p. 6.
53 Guy Anderson and Jon Grevatt, "Briefing: A Market Set to Make its Mark", Jane's Defense Weekly, 27 May 2011 (http://www.navyrecognition.com/index.php?option=com_content&task=view&id=233).
54 *SIPRI Military Expenditure Database* (http://www.sipri.org/research/armaments/milex/milex_database).
55 Balakrishnan and Matthews, "The Role of Offsets in Malaysian Defense Industrialization", op. cit., pp. 345–350, 354–356; *Revitalizing Indonesia's Defense Industrial Base*, op. cit., pp. 7–8.
56 *IHS Jane's Navigating the Emerging Markets: Malaysia* (London: IHS Global Ltd, January 2012), p. 11.
57 *Indonesia's Emerging Defense Economy: The Defense Industry Law and Its Implications*, RSIS Policy Report (Singapore: S.Rajaratnam School of International Studies, August 2013), pp. 4–5.
58 Andrzej Jeziorski, "AIDC Still Aims for Privatization," *Flight International*, August 25, 1999, p. 23.
59 "Taiwan: Industry Plays Vital Part in Modernization," *Jane's Defense Weekly*, July 8, 1998; Kelly Her, "Military Producer of IDFs Changing into Private Firm," *Free China Journal*, October 20, 1995, p. 3.
60 "Taiwan to Privatize Lone Aircraft Maker AIDC," *AFP*, March 19, 2013.
61 Chris Pocock, "AIDC Aims for Greater Role as Aerospace Sector Supplier," *AIN Online*, January 28, 2010 (http://www.ainonline.com/aviation-news/defense/2010-01-28/aidc-aims-greater-role-aerospace-sector-supplier); AIDC website (www.aidc.com.tw).
62 AIDC website (www.aidc.com.tw); Globalsecurity.org, "Aerospace Industry Development Corporation (AIDC)," (http://www.globalsecurity.org/military/world/taiwan/aidc.htm).
63 CSBC website (http://www.csbcnet.com.tw).
64 Jon Grevatt, "Dearth of Engineers to Slow Indonesia's Defense and Industrial Development", *Jane's Defense Industry*, 29 July 2011.
65 Grevatt, "Indonesian Aerospace Transport Aircraft Contract Nears Government Approval", op. cit.; Dzirhan Mahadzir, Alex Pape and Craig Caffrey, "Facing the Future: Malaysian Armed Forces", *Jane's Defense Weekly*, 7 April 2008.
66 Kwang-il Baek and Chung-in Moon, "Technological Dependence, Supplier Control and Strategies for Recipient Autonomy: The Case of South Korea", in *The Dilemma of Third World Defense Industries* edited by Kwang-il Baek, Ronald. D. McLaurin, and Chung-in Moon, (Boulder, CO: Westview Press, 1989), p. 157.
67 Juergen Brauer, "Arms Production in Developing Nations: The Relation to Industrial Structure, Industrial Diversification, and Human Capital Formation", *Defense Economics*, April 1991, p. 166.
68 Cheng and Chinworth, "The Teeth of the Little Tigers: Offsets, Defense Production and Economic Development in South Korea and Taiwan", op. cit., pp. 275–276.
69 See Kogila Balakrishnan and Richard A. Bitzinger, *ASEAN Defense Industrial Collaboration: Getting to Yes*, S.Rajaratnam School of International Studies, RSIS Commentary No. 232, 21 December 2012 (https://www.rsis.edu.sg/wp-content/uploads/2014/07/CO12232.pdf).
70 Ibid.

7 Conclusions
The future of military technonationalism in Asia

As a strategy for Asian defense industrialization, military technonationalism has been an arguable success. Before World War II most Asia-Pacific nations possessed few or no means for indigenous arms manufacturing, and the one country that did – Japan – saw that capacity either destroyed in the war or largely dismantled afterwards. Starting basically from scratch in the 1950s and 1960s, China, India, Japan, and South Korea were able construct (or reconstitute) substantial defense industries, and able to produce a broad range of defense materiel. Even Indonesia was able to achieve a modicum of self-reliance in arms acquisition in certain sectors, such as aerospace, as were other countries in the region, including Singapore and Taiwan. On the surface, therefore, a state-led national defense innovation and manufacturing strategy, driven to a large extent by the argument that autarky is a crucial element of national security, appeared to pay considerable dividends.

Impelled by military technonationalism, many countries in the Asia-Pacific have created extensive, even quite impressive, local arms industries. In some areas these countries are moving toward the point where they are capable of producing arms that approach the global state of the art. South Korea, for example, manufactures an advanced trainer jet (the T-50), and its K9 self-propelled artillery and K2 tank are likely as capable as any comparable systems produced in the West or Russia. India's *Tejas* fighter is impressive in its extensive use of carbon fiber composites, which make up 45 percent of the plane's airframe by weight, including the fuselage, wing, elevons, and vertical stabilizer. Taiwan has carved out an no niche in building many types of tactical missile systems – in particular, even developing a supersonic anti-ship cruise missile – while Singaporean weapons have come to earn a reputation for quality and reliability, reflected in the country's recent arms-export successes (even selling to Western countries, e.g., exporting the Bronco all-terrain tracked vehicle to the British army). Not surprisingly, Japan, as an advanced industrial nation, manufactures some very advanced weapons systems, particularly in the area of submarines, fighter aircraft, main battle tanks, and missile systems.

China in particular has made significant progress in developing a modern, world-class defense industrial base. For more than 20 years China been engaged in a concerted effort to modernize and upgrade its arms industry, and over the past decade this has begun to produce significant results, as evidenced in the growing numbers of new types of Chinese weapons that are increasingly comparable to Western systems in terms of quality and capabilities. These include the J-10 fighter, the *Yuan*-class diesel-electric submarine, the Type-052C/D destroyers, and the HQ-9 long-range surface-to-air missile. In addition, Beijing has embarked on an array of new military research and development (R&D) projects, including two fifth-generation combat aircraft programs (the J-20 and J-31), the DF-21D anti-ship ballistic missile (ASBM), new nuclear *and* conventionally powered submarines, hypersonic weapons, and an indigenous aircraft carrier. Overall, the quality and quantity of the Chinese military, *based almost exclusively on indigenously produced weapons systems*, has been expanding at a steady rate for over two decades, and this growth in military capabilities will likely continue for at least another decade or more. The rejuvenation and expansion of the Chinese defense technological and industrial base has been backed up by a long-term, top-down commitment to provide it with the resources it needs to support such an undertaking. In 2016, for example, China allotted around US$146 billion to defense, making it the second-highest military spender in the world. Of this, approximately one-third (approximately US$48 billion) was allocated for weapons development and acquisition; in turn, a conservative estimate would peg the Chinese military R&D budget at around US$10 billion, at least, outstripping R&D spending by every other country save the United States.

But is it acceptable to say that the technonationalist approach has been an effective and efficient strategy for armaments production? Armaments production in the Asia-Pacific, in terms of innovation, product, and sales, continues to run a poor third to the traditional leading arms-producing states, e.g., the United States, Russia, and Western Europe – and even to defense industries in newly industrialized states in other parts of the world, such as Israel and Brazil. Overall, most defense industries in the region are still primarily platform-centric "metal-bashers" as opposed to innovators. Most weapons systems produced in the Asia-Pacific, while good, are still rather prosaic and "industrial-age": tanks, artillery pieces, surface combatants, combat aircraft, etc. To be sure, the Asian arms industry has produced a few interesting, even cutting-edge military systems – for example, South Korea has developed its own anti-ship and land-attack cruise missiles, China has built the world's only ASBM and is also working on hypersonic delivery systems, and most large arms-producing states in the region have ambitions to develop fifth-generation fighter jets. In general, however, most defense industrial bases in the Asia-Pacific have historically been rather weak or disappointing performers. Countries like Indonesia and Taiwan have experienced considerable setbacks in their efforts to manufacture their own armaments and have been forced to dramatically scale back their expectations. The 1997–98 financial crisis forced South Korea to confront serious overcapacities in its domestic arms industry as part of its broader campaign to reform and rein in its bloated and already over-extended *chaebols* (industrial conglomerates). Beginning in the late 1990s Seoul initiated a series of initiatives to "right-size" the national defense industrial base. The most significant outcome of this rationalization effort was the formation in late 1999 of Korea Aerospace Industries (KAI), the result of a forced merger of three of the country's four aircraft companies.

India has been a particularly disheartening case study. After China, India possesses the largest and most ambitious defense industrial base in the Asia-Pacific, and yet its performance over the past 50 years has been nothing short of disappointing. New Delhi has squandered billions of dollars on domestic weapons programs that have never lived up to their promises when it came to capabilities, costs, or deliveries. Overall, the Indian defense technological and industrial base remains a bloated white elephant of highly protected, monopolistic state-owned corporations, headed by an unresponsive and arrogant DRDO, which presses for indigenous production while paying little heed to requirements or program milestones. Reforms have been mooted, including opening up much of the defense bidding process to private companies, but the current Modi government will have its work cut out for it.

Even Japan, arguably the most advanced defense industrial base in the Asia-Pacific, has been failed to perform as of late. The Japanese defense industry has been described as a "boutique" business, supplying limited numbers of hand-crafted (if high-quality) armaments to a single customer (i.e., the Japan Self-Defense Force, or SDF).[1] In return, the government has ensured, through SDF procurement, guaranteed production work, profitability, and *de facto* subsidies for R&D.[2] Lately, however, the structural defects of this rather cozy system, coupled with rising the Japanese development and manufacturing costs over a decade of stagnation and neglect in the defense budget, and the traditional ban on arms exports have taken its toll on the *kokusanka* concept.[3] As Yukari Kubota has put it so succinctly, Japan's "traditional defense business model, in which the government relies on a contractor for R&D and production while the contractor recovers its prior investment through mass production in the close public-private relationship, is no longer functioning well."[4]

"No longer functioning well" is putting it mildly. Simply put, Japan's technonationalist defense industrial paradigm has become increasingly untenable, and Tokyo is facing a "Sophie's choice" of which defense industrial sectors it can sustain and which it has to let go of. To a certain extent the decision is already being made for the government, as some of industry is "voting with its feet" by exiting the defense industry. A 2009 study by Japan's Ministry of Defense showed that 13 companies that used to manufacture equipment or components for the Ground Self-Defense Force (GSDF) have gone bankrupt since 2003. A further 35 firms working for the GSDF have simply exited the defense business. In addition, 20 other companies engaged as subcontractors to the country's fighter jet industry have either withdrawn from this business or plan to do so, including Sumitomo Electric, the country's sole producer of nosecones for Japanese fighter aircraft.[5]

One result of this dysfunction was the 2014 decision by the Abe government to overturn Japan's 50-year-old ban on arms exports. Even then, overseas arms sales are hardly assured, as was evidenced by Japan's failure in early 2016 to seal a lucrative deal to export submarines to Australia (the contract instead went to a French shipbuilder). In general, Japan is likely to discover what other Asian-Pacific arms producers have already learned: given a well-saturated and intensely competitive global arms market, and the presence of several large, already established arms exporters (such as the United States, Russia, Western Europe, and Israel), securing overseas defense sales is going to be much harder than it looks.

Only China, it seems, has the potential to live up to its technonationalist goal of creating a large, modern, full-spectrum defense technology and industrial base that is technologically competitive on a global scale. In this regard it has benefitted from a number of developments. In the first place, modernizing the Chinese defense industry and upgrading its level of technology has long been a first-order priority of the Chinese communist regime. As such, considerable sums of money have been spent on improving and expanding military R&D, on reforming, restructuring, and renovating domestic production facilities, and on strengthening central government oversight and control of the whole process of military research, development, production, acquisition, and deployment. Additionally, Beijing has for more than 20 years committed itself to large annual increases in defense spending – including procurement spending – resulting in large orders for local armaments factories. Finally, the government has greatly encouraged and facilitated overseas arms sales. China, in fact, consistently ranks among the top arms-exporting nations according to the Stockholm International Peace Research Institute (SIPRI); during the period 2011–15, in fact, China was placed number three, capturing nearly 6 percent of the global arms market and chalking up sales worth US$8.4 billion.[6]

The failure of military technonationalism

Upon closer inspection it becomes apparent that the technonationalist approach to armaments production has many drawbacks and shortcomings. In the first place the technonationalist approach has often resulted in a considerable amount of effort and resources being wasted on "reinventing the wheel," i.e., duplicating weapons systems that are widely and more cheaply available on the global arms market, such as main battle tanks, armored vehicles, trainer aircraft, air-to-air missiles, and even small arms (e.g., Singapore manufactures its own assault rifle). Quite often, in fact, Asian armaments programs are chosen more for their "doability" (i.e., the likelihood that they will succeed) and for their potential to support and advance local arms industries rather than for their actual military need. In these cases military requirements too often become procrustean things, therefore stretched to fit to what local industries can deliver rather than driving acquisition, R&D, and production. As Richard Samuels has put it so bluntly, "in Japan, by far the most important thing about a weapon is learning how to make it."[7]

In addition, locally produced armaments are frequently acquired not for their capabilities but for reasons of economics, i.e., to provide employment and to keep factories operating (for instance, one Japan defense industry expert has argued that Japan has largely given up on the idea of autarky in arms acquisition, and instead Tokyo now runs most of its domestic weapons projects mainly as "jobs programs"[8]). At the same time local arms manufacturers tend to push their governments to buy those systems that they are already capable of producing, i.e., "legacy systems," which only compounds the problem of expensive replication. Even then the economic argument is specious: Defense industries actually employ very small numbers of workers, many of which could be shifted to non-military production if necessary. At the same time, it is almost always less expensive to import finished weapons systems than it is to develop and manufacture them domestically (which often entails the creation of whole new industries). As the *Economist* has noted, "In a market economy, the whole concept of strategic industries is suspect."[9]

As an industrial strategy, too, military technonationalism has generally been a very expensive endeavor, especially when compared to its often-modest results. Japan, for example, makes some of the most costly weapons in the world, due in large part to small, extended production runs. Its indigenous F-2 fighter jet, for example, has a price tag of least US$120 million apiece, or approximately three times that of the US F-16 combat plane upon which it is based. Moreover, escalating expenses caused the Japanese to cut total F-2 production from 141 to only 94 planes, which only further increased its unit price.[10] As Christopher Hughes puts it: "Japan's defense planners have sought the 'holy grail' of pure indigenous defense production and consequent technological autonomy . . . even if these present development risks and high procurement costs."[11]

In addition, low levels of defense spending have only compounded the region's problems achieving efficient production; most countries in the region cannot afford high enough levels of annual procurement so as to make localized production cost effective. Japan's Type-10 main battle tank, for example, costs over US$11 million apiece, compared to around US$8.5 million for a comparable U.S.-built M-1A2 Abrams tank, in part because it is produced at a rate of just a dozen or so per year. At the same time, except for China, most Asian-Pacific countries actually spend very little on defense R&D, making it difficult to engage in meaningful innovation in the first place, or else forcing countries to be highly selective when it comes to R&D programs – which in turn can often drive a country to stick to areas of R&D where it has pre-existing experience and expertise, even if those sectors are already saturated with competitive systems (again, tanks, armored vehicles, small arms, etc.). Meanwhile, in some cases – particularly India – R&D funds are frequently lost through waste, abuse, and corruption, resulting in program delays, cost overruns, and weapons systems that never meet their required performances.

Most domestically produced Asian armaments are hardly that much of an improvement over the foreign weapons they are intended to supplant. In fact, most of the time better and cheaper military equipment can be found on the global arms market. Hughes argues that Japan's indigenous F-1 fighter, built during the late 1970s/early 1980s, was "obsolete almost as soon as it went into production."[12] For its part, the Indian military has been forced to continually scrounge for foreign stopgaps to compensate for underperforming local systems, buying Su-30MKI combat aircraft and T-90 tanks from Russia and Rafale fighter jets from France to make up for delays and problems with such homegrown weapons systems as the *Arjun* tank or the *Tejas* LCA. Even the licensed-production of a foreign weapons system has rarely been cost effective. According to a study cited in the *Economist*, defense offsets typically raise the price of arms procurement by around 20 to 30 percent when compared to buying off the shelf.[13] For example, Seoul's insistence on locally manufacturing the US F-16 fighter added an estimated 20 percent to the total cost of the program. As Michael Green put it, "expensive domestic development programs," came at a "heavy price to pay in terms of military efficiency."[14] Although he was referring only to Japan, this assertion could be applied to most arms-manufacturing activities in Asia.

In addition, indigenous defense industries and local armaments programs have been highly prone to failure. The story of India's defense industry, for instance, is a nearly unbroken story of disappointments and setbacks. The country's indigenous *Tejas* fighter jet, initiated in the mid-1980s, did not have its first flight until 2001, more than a decade behind schedule, and it only achieved full operational clearance with the Indian Air Force in 2015. India's *Arjun* main battle tank did not enter service with the Indian Army until 2011, more than 30 years after the program was initiated. The program has had a history of technical problems, resulting in horrendous delays and cost overruns: the tank is reportedly more than 16 years behind schedule and 20 times over its original cost estimates.[15] For its part, Indonesia's aerospace industry never found any traction, failing to find even one viable long-term aircraft-manufacturing program. In the late 1990s, after pouring billions of dollars into various government-initiated aircraft programs – including an unsuccessful effort to build a 50-seat turboprop commercial airliner – Jakarta finally pulled the plug on the company; the state-run aircraft company IPTN was restructured, and at least three-quarters of the workforce was laid off.[16] In the cases of Japan and Korea, local arms industries still require considerable financial support from their governments in the form of preferential supply contracts or excess production runs; in 2000, for example, the Korean government decided to buy additional F-16 fighters – over the objections of its air force, which claimed that they were unnecessary – in order to keep KAI, one of the country's key aerospace firms, open until the T-50 went into production.[17]

Finally, history has shown that overly autarkic and "introverted" armaments producers tend to lag behind the first-tier arms-producing states when it comes to innovation. As Vernon and Kapstein argued more than a quarter-century ago, "any nation that is determined to rely upon its own products, its own technologies, and its own enterprises to fulfill its defense needs will pay a far higher premium for such a policy . . . costs that will be expressed not only in terms of money but also in a sacrifice in the quality of its military equipment."[18]

Consequently, military technonationalism comes with some very high opportunity costs: countries pay a huge price for limited autarky in armaments acquisition while also risking losing access to global dynamics of technology diffusion and innovation. Ultimately, therefore, technonationalism may be just was Reich was arguing in the beginning: that it is more an emotional appeal based on "irrational nationalism" than it is a sound strategy for armaments production.[19]

The way forward for Asian arms producers: go big, go high-tech, or go home

So how can one assess the success or failure of military technonationalism in Asia, then? In the first place it has certainly been an important, perhaps even primary, driver of defense industrialization in the region; it has motivated several countries in the Asia-Pacific to pursue arms production even when the realities of economics or technology acquisition failed. South Korea, for example, is arguably too small (in terms of its defense R&D base, its defense budget, and national arms market) to justify it pursuit of such an ambitious defense industrialization strategy, and yet powerful social, historical, and political dynamics have made it continue, despite all the evidence that it should not even try. Japan, the grandfather of the technonationalism, is presently facing of a crisis of confidence when it comes to its historical approach to arms manufacturing; yet even that has not prevented Tokyo from pursuing the development of a fifth-generation fighter jet (i.e., the ATD-X).

On the other hand, in many instances it is apparent that the technonationalist model is failing. Armaments production is a costly endeavor; it cannot be done "on the cheap." And as countries such as India and Indonesia have shown, technonationalism is no substitute for a weak R&D base – indeed, just the opposite is true, as the process of technology acquisition, diffusion, and nurturing requires constant care and feeding. Technonationalism also does not compensate for the lack of sizable arms markets. In fact, in the drive for autarky in armaments production countries have too often glossed over the need for large enough production runs in order to ensure cost-effectiveness; much of the time such concerns are waved away with unrealistic assumptions of huge overseas sales. (In the early 2000s, for example, the Korea Aerospace Industries persistently asserted that it expected to export 600 to 800 of its T-50 advanced jet trainers; in fact, it has so far sold only 56 planes to overseas customers.)[20]

So what are the possible choices facing Asian armaments producers? In the first place, countries can continue to commit themselves to pursuing technonationalist aims and approaches. This, however, will demand that their governments and industries also commit themselves, over the long term, to adequately and consistently underwriting such an endeavor – in other words, providing sufficient funding for R&D, guaranteeing new defense projects, ensuring large enough procurement orders so as to make production cost effective, keeping factories properly run and modern, and expanding efforts to find overseas markets for their arms. Even then, these efforts may not assure success, as the global struggle to develop and produce weapons is highly competitive and dominated by a few large players. In fact, of all the countries in the Asia-Pacific region, perhaps only China has the potential to succeed as a broad-based arms producer given its large domestic arms market and sizable and steadily increasing defense budget.

On the other hand, countries could choose to jettison the more under-producing and unprofitable sectors of their arms industries and instead concentrate on niche areas where they possess demonstrable expertise. Such a strategy should also be explicitly linked to the export market, and countries should especially focus on defense products and services where the prospects for overseas sales are more likely. In this regard they should attempt to emulate such high-tech niche suppliers as Israel (drones and electro-optics) and Brazil (ground attack aircraft), and try to carve out lucrative nooks for themselves in the global arms marketplace. Singapore, for example, is attempting this with its armored vehicles businesses, and South Korea has had some success as an exporter of amphibious assault ships and submarines; Japan might find a small but lucrative market for its unique US-2 flying boat or its C-2 transport plane.

Even such a high-tech, export-oriented niche production strategy is still a risky strategy, as the global arms market is pretty competitive in nearly every segment. It would also require these countries to abandon any ideas of achieving across-the-board autarky (even as a long-range goal) and accept a much more limited version of an indigenous defense industry. This "limited autarky" could still preserve at least some capacity for indigenous production, partly for economic reasons (i.e., maintaining existing defense industrial bases, protecting jobs, balance of payments, arms exports, etc.). It could also provide at least "strategic sovereignty" in that armaments production could serve as a hedge against arms embargoes or supplier restrictions. In addition, having at least some indigenous technological-industrial capabilities permits a nation to develop customized solutions for national defense, as well as encourages a sense of empowerment and self-confidence in that it can domestically produce at least *something* that contributes to its national security.[21] At the same time, however, such an approach would only be practical if the product or products to be acquired were deemed so strategically vital as to make import-dependency politically and militarily unacceptable, and if they could be manufactured in a cost-effective manner. Such parameters greatly limit the number of likely systems or products that could conceivably be domestically sourced. Even "limited autarky," therefore, might make little sense.

Finally, of course, these countries could simply decide to remain at the low-tech end of arms manufacturing, e.g., constructing hulls for naval vessels, small arms production, licensed assembly of imported weapons kits, etc. Even then, such a modest approach to arms production – a low-budget form of "limited autarky," as it were – could still be a challenge, as it will be continually *ad hoc* and difficult to make economically sustainable. More importantly, such low-end armaments production would not make much of a contribution to overall national economic and technological development, nor would it help to create a foundation for high-tech defense industries.

In the end, when it comes to armaments production the Asian-Pacific nations are being given a hard lesson: go big, go high-tech (particularly when it comes to export-oriented niche production), or go home. The question is, will these countries learn and adapt? If technonationalism continues to drive these countries' defense industrial strategies, then the answer, sadly, is no.

Notes

1 International Institute for Strategic Studies (IISS), "Current Trends in Asia-Pacific Defense Industries," *The Military Balance 2009* (London: Routledge, 2009), p. 455.
2 Hughes, *The Slow Death of Japanese Techno-Nationalism?*, pp.11–12.
3 Paul Kallender-Umezu, "Japan Strives to Overcome Defense Industrial Base 'Crisis,'" *Defense News*, June 24, 2012.
4 Yukari Kubota, "Japan's Defense Industrial Base in Danger of Collapse," *AJISS-Commentary*, May 10, 2010 (http://wsn.globalo.com/Japan/Kubota-Yukari/Japans-Defense-Industrial-Base-in-Danger-of-Collapse).
5 Hughes, *The Slow Death of Japanese Techno-Nationalism?*, p. 30; Jon Grevatt, "Briefing: Spending Dearth Withers Japan's Defense Industrial Base," *Jane's Defense Weekly*, October 1, 2009.
6 Aude Fleurant, et al., *Trends in International Arms Transfers, 2015*, SIPRI Fact Sheet, February 2016, p. 3.
7 "Survey of the Global Defense Industry: Home Alone," *Economist*, June 14, 1997, p. 10.
8 Author's interview with Michael Green, February 2011.
9 Iain Carson, "A Survey of the Defense Industry: Odd Industry Out," *Economist*, July 20, 2002, p. 6.
10 Hughes, "The Slow Death of Japanese Techno-Nationalism?" p. 459.
11 Hughes, "The Slow Death of Japanese Techno-Nationalism?" pp. 454–455.
12 Hughes, "The Slow Death of Japanese Techno-Nationalism?", p. 459.
13 "Guns and Sugar: The Defense Industry," *Economist*, May 25, 2013.
14 Green, *Arming Japan*, p. 4.
15 Shiv Aroor and Amitav Ranjan, "Arjun, Main Battle Tanked," *Indian Express*, November 14, 2006; Laxman Kumar Behera, "The Saga of MBT-Arjun," *Defense Review Asia*, June 2010, pp. 20–22.
16 Margot Cohen, "New Flight Plan," p. 45; *IHS Jane's Navigating the Emerging Markets: Indonesia*, p. 21.
17 "Air Force Protests Decision to Produce Older Jet Fighters," *Korea Times*, May 12, 1999.
18 Raymond Vernon and Ethan B. Kapstein, "National Needs, Global Resources," *Daedalus* (Fall 1991), p. 19.
19 Green, *Arming Japan*, pp. 11–12.
20 Philip Finnegan, "S. Korea's New Trainer Jet Faces Tough Market," *Defense News*, March 13, 2000; Jun Ji-hye, "KAI will Export T-50s to Thailand," *Korea Times*, September 17, 2015.
21 See Bitzinger, *New Ways of Thinking about the Global Arms Industry*, pp. 6–7.

Index